T0305739

Bayesian Mediation Analysis using R

Delve into the realm of statistical methodology for mediation analysis with a Bayesian perspective in high dimensional data through this comprehensive guide. Focused on various forms of time-to-event data methodologies, this book helps readers master the application of Bayesian mediation analysis using R. Across ten chapters, this book explores concepts of mediation analysis, survival analysis, accelerated failure time modeling, longitudinal data analysis, and competing risk modeling. Each chapter progressively unravels intricate topics, from the foundations of Bayesian approaches to advanced techniques like variable selection, bivariate survival models, and Dirichlet process priors. With practical examples and step-by-step guidance, this book empowers readers to navigate the intricate landscape of high-dimensional data analysis, fostering a deep understanding of its applications and significance in diverse fields.

Dr. Atanu Bhattacharjee is a academic statistician at the School of Medicine, University of Dundee. He is an expert in the field of medical statistics, with a focus on survival analysis, competing risks, and high-dimensional data.

Dr. Bhattacharjee's research interests include the development of new statistical methods for the analysis of time-to-event data, with a focus on the analysis of competing risks and high-dimensional data. He has published several research papers and articles in leading statistical journals on these topics.

Dr. Bhattacharjee has also contributed to the development of R package, which can be used to perform competing risks analysis and high-dimensional data analysis respectively.

Bayesian Mediation Analysis using R

Atanu Bhattacharjee

CRC Press
Taylor & Francis Group
Boca Raton London New York

CRC Press is an imprint of the
Taylor & Francis Group, an **informa** business

A CHAPMAN & HALL BOOK

First edition published 2024
by CRC Press
2385 NW Executive Center Drive, Suite 320, Boca Raton FL 33431

and by CRC Press
4 Park Square, Milton Park, Abingdon, Oxon, OX14 4RN

CRC Press is an imprint of Taylor & Francis Group, LLC

© 2024 Atanu Bhattacharjee

Library of Congress Cataloging-in-Publication Data

Names: Bhattacharjee, Atanu (Statistician), author.
Title: Bayesian mediation analysis using R / Atanu Bhattacharjee.
Description: First edition. | Boca Raton : C&H/CRC Press, 2024. | Includes
bibliographical references and index.
Identifiers: LCCN 2023048313 (print) | LCCN 2023048314 (ebook) | ISBN
9781032287508 (hbk) | ISBN 9781032287522 (pbk) | ISBN 9781003298373
(ebk)
Subjects: LCSH: Mediation (Statistics) | Bayesian statistical decision
theory. | R (Computer program language)
Classification: LCC QA278.2 .B498 2024 (print) | LCC QA278.2 (ebook) |
DDC 519.5/42--dc23/eng/20231120
LC record available at https://lccn.loc.gov/2023048313
LC ebook record available at https://lccn.loc.gov/2023048314

ISBN: 978-1-032-28750-8 (hbk)
ISBN: 978-1-032-28752-2 (pbk)
ISBN: 978-1-003-29837-3 (ebk)

DOI: 10.1201/9781003298373

Typeset in CMR10
by KnowledgeWorks Global Ltd.

Publisher's note: This book has been prepared from camera-ready copy provided by the authors.

This book is dedicated to my beloved babu, ma, Sumita, and Biplab.

Contents

Preface

The goal of this book is to provide a comprehensive guide to statistical methodology for mediation analysis using a Bayesian approach in high-dimensional data, specifically focusing on different types of time-to-event data methodologies. The purpose is to encourage the use of high-dimensional data analysis in everyday practice. The book is divided into ten chapters, each delving into the application of Bayesian mediation analysis in high-dimensional time-to-event data using R. Chapter 1 introduces the concepts and definitions of mediation analysis and provides an overview of its application using R. Subsequent chapters cover topics such as single, multiple, path, and conditional process of mediation analysis.

In chapter 2, the Bayesian approach to mediation analysis will be described, including statistical inference and variable selection. The chapter will also cover topics such as prior and posterior distributions, model building procedures, and assessment using R. Examples using R will be provided to demonstrate the application of mediation analysis.

Chapter 3 will cover the topic of survival analysis, explaining the concepts and semiparametric approaches. It will also introduce the Bayesian approach to working with survival analysis and provide examples of different methods such as Cox proportional hazard assumption and parametric survival analysis with high dimensional data. The chapter will demonstrate the statistical methodology for different survival models using R. Chapter 4 will delve into the issue of competing risk in oncology data, specifically when a cancer patient dies from causes other than cancer. The decision-making process about biomarkers in the presence of competing risks is often challenging, and this chapter will present a Bayesian approach to analyzing competing risks using R.

In chapter 5, the topic of accelerated failure time (AFT) modeling will be discussed in-depth. The author will present an application of AFT as an alternative to Cox proportional hazard models. The chapter will cover graphical methods, model assessment, inference for log-location scale, and semiparametric multiplicative hazard models. The application of AFT models using R will also be illustrated with examples.

In chapter 6, the concept and definition of longitudinal data analysis will be discussed, and the R software will be used to analyze repeatedly measured data. Different techniques such as mixed-effect model, mixed-effect polynomial, covariance structure, generalized estimating equation, and missing data handling methodology will be presented. Additionally, variable selection techniques will be explained, with a focus on high-dimensional data and illustrated using R.

Chapter 7 will focus on the concept of high-dimensional data and its importance in extensive data analysis. High-dimensional data measured with time-to-event information will be illustrated, and variable selection techniques in high-dimensional data such as Zellner's prior, stochastic variable selection, and Gibbs sampling variable selection will be detailed. Examples using R will be provided to demonstrate the application of high-dimensional data analysis.

In chapter 8, the topic of Bayesian survival mediation analysis using R will be discussed. The chapter will cover statistical methodology with bivariate survival models, smoothing techniques, nonparametric Bayesian methods, and Dirichlet process priors, with a focus on the application of Bayesian mediation analysis in survival data.

Chapter 9 will focus on Bayesian Accelerated Failure time mediation analysis, discussing AFT in high-dimensional survival data and Bayesian AFT in high-dimensional data. The chapter will provide a step-by-step illustration of different survival analysis methodologies in high-dimensional mediation analysis using R.

Finally, chapter 10 will cover the topic of competing risk modeling, specifically discussing the Bayesian approach with high-dimensional data analysis. The chapter will present the definition, concept, and importance of competing risks in oncology data, and the explained methodology will be illustrated using R data examples.

Author

Dr. Atanu Bhattacharjee serves as an Academic Statistician at the University of Dundee, Scotland, specializing in medical statistics. His expertise encompasses survival analysis, competing risks, and high-dimensional data analysis.

Dr. Bhattacharjee's research revolves around advancing statistical methodologies for analyzing time-to-event data, particularly emphasizing competing risks and high-dimensional data. His contributions are evident through numerous publications in esteemed statistical journals.

Additionally, Dr. Bhattacharjee has played a pivotal role in developing an R package tailored for conducting competing risks analysis and high dimensional data analysis.

Chapter 1

Mediation Analysis

1.1   Introduction

Mediation analysis stands to compare the relationship between two variables, i.e., X and Y, while X is presented as a possible cause of Y. It incorporates the randomization of units of the Y along with the independent unit value of X. Mediation analysis provides the scope to include the third variable in the relation of X with Y, where the mediator variable is M that causes X. It is formulated as X → M → Y. Now, Mediation gives the scope to incorporate several variables as presenting the third variable Z. The term Z can cause both X and Y, and it is a suitable example of a confounding variable. Sometimes, Z relates to Y and X. It improves the prediction of Y by X but does not substantially relate to X to Y while Z is included in the analysis. Now Z improves the relation of X to Y by incorporating it in the analysis. Sometimes, Z may improve the link of X on Y like the connection differs at different values of Z. We may modify the consideration of X to Y by the association of X to Y as it differs to varying values of Z. It is an example of the moderator or interaction effect.

The statistical technique is used to compute the relationship between the pre-existing variable causing a mediation is called Mediation analysis [1]. A mediation model help to recognize and explain the procedure, which shows a relationship among the independent and the dependent variable and one additional variable known to be the mediator variable. This third hypothetical variable is also called an intermediary or intervening variable [2]. Figure 1.1 shows the relationship between all three of them. The aim is to release the effects of a treatment on an outcome through alternative causal mechanisms. It established both continuous and discrete products and was extensively used in research fields.

However, it is commonly adopted in clinical research, psychology, and applied research [2]. This application became very uncommon while we explored the application with time-to-event data analysis. Most commonly, the time to event data analysis is explored by the additive hazard model, accelerated failure time, or proportional hazard model. However, the causal mediation analysis becomes useful to work with time to event data [3]. The usage of the product coefficient method proves a valid test to conclude the presence

DOI: 10.1201/9781003298373-1

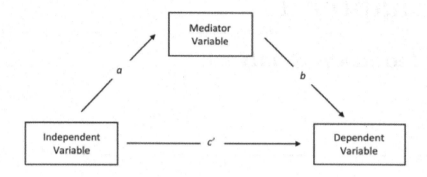

FIGURE 1.1: Simple mediation model.

of mediator effect with proportional hazard model, but it does not provide any measurement. Avin [4] introduced path-specific effects (PSEs) to generalize mediation analysis from a single mediator to multiple mediators. The traditional approach of this method for survival data compares the changes of hazard ratios of exposure between the Cox models with and without adjusting for the potential mediator. Different possible effect decomposition is studied for the survival function, hazard, mean survival time and median survival time, and median survival scales [5]. The mediation analysis is approached by the (I) Casual Steps [1], (II) Testing mediation effect by (A) Sobel test [6] (B) Correction to Aroian test, (C) Distribution of Products, and (D) Bootstrap analysis [7].

One of the widely used statistical tests in Mediation analysis is the Sobel test. It is used to test the significance of the mediation effect [8]. It helps to identify if the reduction in the effect of the independent variable after the involvement of the mediator variable is significant and hence whether the mediation effect is stochastically significant. It is one type of t-test. The α term denotes the magnitude of the relationship between the independent and mediator variable. Similarly, the β shows the magnitude of the relationship between the mediator and dependent variable after controlling the effect of an independent variable. Thus, the product of these two terms represents the amount of variance in the dependent variable, which accounts for the independent variable of the mediator. The Sobel test applies the magnitude of the indirect effect compared to its estimated standard error of measurement to derive a t statistic [6].

$$t = \frac{\tau - \tau'}{SE} \text{ or } t = \frac{(\alpha\beta)}{SE} \tag{1.1}$$

where SE is the pooled standard error term and

$$SE = \sqrt{\alpha^2 + \beta^2\sigma^2\alpha} \tag{1.2}$$

and σ^2 is the variance of β and $\sigma^2\alpha$ is the variance of α. Sometimes it becomes challenging to work with Mediation analysis with high-dimensional data. High-dimensional data is defined as a dataset in which the number of features p is larger than the number of observations N. Notationally, $p >> N$ [9]. This data structures are studied in many medical research fields having genetic data, imaging data, health outcomes, clinical data, etc. A large number of variables and relatively few subjects resemble such type of data. The aim of the analysis is to detect various patterns within the overall dataset. There are two perspectives to study mediation analysis, namely statistical and casual. Among these two, statistical mediation approaches a regression model to estimate the strength of intervention mediator and mediator outcome effects. Simple mediation analysis emphasizes sensitivity analyses to examine the robustness of the impact.

1.2   Single Mediator Model

A single mediator model is a statistical model that is used to understand the relationship between an independent variable, a mediator variable, and a dependent variable. The independent variable is the variable that is believed to have an effect on the dependent variable, the mediator variable is the variable that explains how the independent variable is related to the dependent variable, and the dependent variable is the outcome of interest. The single mediator model is used to examine the direct and indirect effects of the independent variable on the dependent variable through the mediator variable. This model can be useful in fields such as psychology, sociology, and epidemiology to understand the underlying mechanisms that link different variables.

1.3   Multiple Mediator Model

A multiple mediator model is a statistical model that is used to understand the relationship between an independent variable, multiple mediator variables, and a dependent variable. The independent variable is the variable that is believed to have an effect on the dependent variable, the mediator variables are the variables that explain how the independent variable is related to the dependent variable, and the dependent variable is the outcome of interest. The multiple mediator model is used to examine the direct and indirect effects of the independent variable on the dependent variable through multiple mediator variables. This model can be useful in fields such as psychology, sociology, and epidemiology to understand the underlying mechanisms that link different variables.

Multiple mediator models can be more complex than single mediator models as they involve multiple mediator variables, which can interact and influence each other. There are different ways to estimate multiple mediator models, such as the product-of-coefficients approach, the decomposition approach, and the causal steps approach. These methods are used to estimate the direct and indirect effects and to test the significance of the mediation effects.

1.4 Causal Mediation Analysis

Causal mediation analysis is a statistical technique used to understand the causal mechanisms through which an exposure affects an outcome. It is used to identify the specific pathways or mechanisms through which an intervention or treatment affects the outcome of interest.

In causal mediation analysis, the effect of an exposure on the outcome is decomposed into two components: the direct effect and the indirect effect. The direct effect represents the effect of the exposure on the outcome that is not mediated by any other variable. The indirect effect represents the effect of the exposure on the outcome that is mediated by one or more intermediate variables.

There are different methods to perform causal mediation analysis, such as the front-door criterion, the backdoor criterion, and the instrumental variable method. All of these methods rely on assumptions about the causal structure of the data and the absence of unmeasured confounding.

Causal mediation analysis is useful in a wide range of fields, including health sciences, social sciences, and marketing. It can help researchers identify the most important mechanisms through which an intervention or treatment affects an outcome, and can inform the development of more effective interventions.

1.5 Path-specific Effects

Path-specific effects (PSEs) are a concept used in causal inference to describe the relationship between an independent variable and a dependent variable through an intermediate variable, also known as a mediator variable. A path-specific effect refers to the effect of the independent variable on the dependent variable through a specific causal pathway, defined by the mediator variable.

In other words, it refers to the effect of a treatment or exposure on an outcome through a specific mechanism or process. For example, in a medical study, a drug may be found to have a path-specific effect (PSE) on blood pressure through its effect on blood vessel dilation.

PSEs can be estimated using various statistical methods such as structural equation modeling, path analysis, and mediation analysis. These methods allow for the estimation of direct and indirect effects of the independent variable on the dependent variable, as well as the proportion of the total effect that is mediated through the mediator variable.

It's important to note that the existence of PSEs implies that the independent variable has different effects on the dependent variable through different pathways, and that these effects may be confounded or modified by other variables. Therefore, it's important to control for these variables in the analysis. In general, standard mediation analysis divides the total effect into two groups of PSEs with only a single mediator. The first is the path of exposures that affects the mediator with respect to previous survival history, and second one is the path of directions that involves the mediator, not with previous survival history. All the mediators are treated as one single block of mediators, and the effect mediated by a specific combination of mediators called the path-specific effect is not always distinguishable [4]. The number of PSEs is 2^k with k mediators, which gradually increases exponentially [10]. PSEs arise if the effect of one variable on another is enumerated with specific casual paths.

1.6   Restricted Mean Survival time

Restricted Mean Survival Time (RMST) is a measure of survival that is used to summarize the survival experience of a population over a specific time interval. It is defined as the mean survival time of the individuals who are still alive at the end of the specified time interval.

RMST is calculated by first computing the survival function, which is the probability that an individual will survive beyond a certain time point. It is then calculated by averaging the survival times of the individuals who are still alive at the end of the specified time interval.

RMST is a useful measure of survival as it provides a summary of the survival experience of a population over a specific time interval, and it is not affected by censoring, which is common in survival analysis. Additionally, RMST can be used to compare the survival experience of different populations or treatments and to evaluate the effectiveness of interventions.

The calculation of RMST can be done using various software such as R, SAS, or STATA. It's important to note that the choice of the time interval is crucial, as it will affect the results and the interpretation of the RMST. The idea of RMST goes back to Irwin [11] and is further implemented in survival analysis by Uno [12]. The local meantime can be a measure of average survival from time 0 to a specific time point. It can be estimated as the area under the curve up to that particular point. In a randomized control trial, RMST was recommended as an addition to the hazard ratio for reporting the effect of an intervention. It is a time-dependent measure and is typically calculated over a defined period with adequate follow-up. It is considered a powerful, robust, and interpretable tool for designing and analyzing clinical studies.

1.7   Confounding and Mediation Analysis

Confounding and mediation analysis are two statistical techniques used to understand the relationship between different variables in a study. Confounding occurs when the relationship between an exposure and an outcome is obscured by the presence of a third variable. For example, in a study of the relationship between smoking and lung cancer, age is a confounder because both smoking and age increase the risk of lung cancer.

Mediation analysis is a method used to identify the mechanisms or pathways through which an exposure affects an outcome. For example, in the smoking and lung cancer example, mediation analysis could be used to identify the biological mechanisms by which smoking causes lung cancer.

Both confounding and mediation analysis are important tools for understanding the relationship between variables and can inform the development of interventions to address a particular health problem.

1.8   Cure Models

Cure models are a type of statistical models used to analyze time-to-event data when some individuals are considered "cured" or "non-event" and are not at risk of experiencing the event of interest. This is often the case in medical research, where some individuals may be cured of a disease and are no longer at risk of experiencing a certain event.

Cure models allow for the modeling of both the event and non-event individuals in the same model. These models typically consist of two components: a hazard function that describes the risk of the event for those individuals who are at risk and a cure fraction, which represents the proportion of individuals who are considered cured or non-event.

There are several types of cure models, such as the proportional hazards cure model, the accelerated failure time cure model, and the mixture cure model. These models can be fitted using various software such as SAS, R, or STATA.

Cure models are particularly useful when the goal of the analysis is to understand the risk of an event occurring in a population that includes both event and non-event individuals, and to estimate the proportion of individuals who are considered cured. They are widely used in fields such as medicine, epidemiology, and engineering to analyze time-to-event data. Cure models are investigated by the Kaplan-Meier method to analyze the cancer survival data since cancer patients are to be the long-term survivors of the disease. The advancement in the treatment and more research on cancer extended the study to cure rates, and hence this can be further used to examine the heterogeneity between the cancer patient. Such models allow us to scrutinize the covariate associated with short and long-term effects [13]. For instance, cure models help to examine if a new therapy is linked with an increase or decrease in long-term survivor probability. There are two classes of cure models -**mixture** and **non-mixture** models. Both of these classes describe short-term and long-term effects. The choice between the two is the study of preference and fit. Mixture cure models: This model involves two types of patients—one who is cured and the other who is not cured. This model is usually with logistic regression. Non-mixture cure models: This include model for patients who are not cured. This model has Weibull and Cox models. Explicitly, a mixture cure model can be written as :

$$\text{Probability alive at time t} = \text{ Probability cured} + \text{Probability not cured} \times$$
$$\text{Probability alive at the time if not cured}$$
$$(1.3)$$

In the case of cured patients, the logistic model is used to study the effects of covariates, and for uncured patients, the distribution of time to event is examined.

1.8.1 Mixture Cure Models

Mixture cure models are a type of statistical models used to analyze time-to-event data when some individuals are considered "cured" or "non-event" and are not at risk of experiencing the event of interest. This type of model allows for the modeling of both the event and non-event individuals in the same model, and for the inclusion of a latent variable, or a latent class, which represents the underlying sub-population of individuals who are at risk and those who are cured.

In mixture cure models, a mixture distribution is used to model the time-to-event data, where the mixture distribution is a combination of two or more subpopulations. Each subpopulation is represented by a different distribution, one for the event individuals and one for the non-event individuals. The latent variable represents the underlying subpopulation to which an individual belongs.

Mixture cure models can be fitted using various software such as SAS, R, or STATA, and can be estimated using maximum likelihood or Bayesian methods. They can also include covariates and allow for the estimation of the proportion of cured individuals and the effect of covariates on the cure rate.

Mixture cure models are particularly useful when the goal of the analysis is to understand the risk of an event occurring in a population that includes both event and non-event individuals and to estimate the proportion of individuals who are considered cured while accounting for the presence of subpopulations with different cure rates. An essential assumption regarding mixture cure models is that the overall survival results of two subgroups from the survival experience are cured patients with its cure fraction denoted as π and uncured patients $(1 - \pi)$. The critical advantage of this model is that it permits covariates to have an influence on cured as well as uncured patients [13]. Mixture cure models can also be interpreted with respect to mortality hazard functions. For example, therapy may increase the proportion of patients who are cured but not affect survival for patients who are not cured. The mixture cure survival model can be presented as:

$$S(t) = \pi SU(t) + 1 - \pi \tag{1.4}$$

where $SU(t)$ is the survival function for uncured patients and π is the probability of not being cured. The effect of covariates on π can be investigated by a logistic link. The effect of covariates on SU(t) can be modeled as follows:

$$SU(t) = SU_0(t)\exp(x'\beta) \tag{1.5}$$

As a result, this model provided two separate conclusions of cured and uncured patients. For cured patients, the effects of covariates on the cure probability are examined through the logistic model, while for uncured patients, the impact of covariates on the distribution of time to event is reviewed.

1.8.2   Non-Mixture Cure Model

Non-mixture cure models are a type of statistical models used to analyze time-to-event data when some individuals are considered "cured" or "non-event" and are not at risk of experiencing the event of interest. These models do not use a latent variable to represent the underlying sub-populations of individuals who are at risk and those who are cured, unlike mixture cure models.

Instead, non-mixture cure models use a single distribution to model the time-to-event data, with a separate parameter to represent the proportion of individuals who are considered cured. These models can be divided into two main categories: parametric and semiparametric.

Parametric cure models assume a specific distribution for the time-to-event variable, such as the exponential, Weibull, or log-normal distributions. These models use maximum likelihood estimation to estimate the model parameters and the proportion of cured individuals.

Semiparametric cure models, also known as non-parametric cure models, do not make any assumptions about the underlying distribution of the time-to-event variable. These models use non-parametric estimation techniques, such as the Kaplan-Meier estimator, to estimate the survival function and the proportion of cured individuals.

Non-mixture cure models can be fitted using various software such as SAS, R, or STATA. They are widely used in fields such as medicine, epidemiology, and engineering to analyze time-to-event data when some individuals are considered cured. In the cancer research study, the non-mixture cure rate model proposed by Yakovlev [14] in 1993 and was further discussed by Chen [15] in 1999 is also said to be bounded cumulative hazard model and promotion time cure model. This model is developed on the basis that the number of cells that remain active after the treatment produces detectable cancer and are likely to be following Poisson distribution. There are a few different advantages, such as its structure being the same as the proportional hazard model, and also it provides clear analysis. This model has a simple structure survival function that can provide a maximum likelihood estimation procedure. In the Cancer study, a non-mixture cure model can be studied for long-term survivors in the data. Let us assume that N be the number of cancer cells after the treatment for a patient. Since the cancer cells proliferate, the number of cancer cells is assumed to follow Poison distribution with mean λ. Let Z_j be the random time for the jth cancer cell to produce a detectable cancer mass. Then, the time to relapse of cancer can be defined by the random variable T such that $T = minZ_j; j = 1, 2, \ldots, N$.

1.9 Cox Proportional Hazard Cure Model

In practical situations, the proportional hazard (PH) cure model is used to study the effects of treatment of uncured patients on the survival distribution. In this case, the PH model is considered for the survival times of uncured patients, whereas logistic distribution is reviewed for the cured patients [16].

Y: indicator variable where an individual experience an event (Y = 1) or never experience event (Y = 0) with Pr(Y=1) = p;

T: time to occurrence of the event when Y=1;

$f(t \mid Y = 1)$: density function of T;

$S(t \mid Y = 1)$: survival function of T;

S(t): marginal survival function of T;

$S(t) = (1 - p) + pS(t \mid Y = 1)$ for $t < m$

Assumption: Independent, non-informative, random-censoring model and that censoring is statistically independent of Y. Farewell [17] used a logistic regression model for the incidence

$$p(z) = P_r(Y = 1; x) = \exp(z'b) \mid (1 + \exp(z'\beta)) \qquad (1.6)$$

where the covariate vector z includes the intercept, and a parametric survival model for $S(t = 1)$. Kuk and Chen generalized this by using a Cox PH model.

$$X(t = 1; z) = X_0(t|Y = 1)\exp(z'\beta) \qquad (1.7)$$

where z is a vector of covariates apart from intercept and $X_0(t = 1)$ is the conditional baseline hazard function. Through b and β, the model is able to separate the covariates effects on the incidence and the latency and, in that sense, provide a flexible class of models when there is a priori belief in a non-susceptible group.The conditional cumulative hazard function is

$$A(t = 1; z) = A_0(t = 1)\exp(z'\beta) \qquad (1.8)$$

where $A_0(t = 1; z) = J; X_0(u = 1)du$. The conditional survival function is

$$s(t = 1; z) = s_0(u = 1) \qquad (1.9)$$

where $S_0(t = 1)$ is the conditional baseline survival function. It is not difficult to see that a mixture of PH functions is no longer proportional and in fact, for a binary covariate, a PH cure model can have marginal survival curves that cross. However, the standard PH model is a special case of a PH cure model in which $P(z)=1$ for all z. The PH cure model is a special case of a multiplicative frailty model, in which the hazard for an individual, conditional on Y, can be written as

$$X(t; z) = Y \times X(t = 1; z) \qquad (1.10)$$

As a frailty variable, Y is not entirely unobservable since an individual becomes labeled as $Y = 1$ if an event is observed.

1.10   Hypothesis in Mediation Analysis

In a causal mediation model, the mediation effect refers to the proportion of the total effect of the exposure on the outcome that is mediated by the mediator. In a high-dimensional setting with continuous mediators, performing a hypothesis test for the mediation effect can be challenging, as the number of mediators may be too large to test each one individually.

One approach for testing the mediation effect in a high-dimensional setting is to use variable selection methods, such as Lasso or Ridge regression, to select a subset of the most important mediators. Once the subset of mediators has been selected, a traditional hypothesis test for the mediation effect can be performed.

Another approach is to use the Bayesian framework, which can handle high-dimensional data naturally. The Bayesian framework allows to incorporate prior knowledge and regularize the model parameters, which can help to identify the relevant mediators in a high-dimensional setting.

In R, there are several packages that can be used to perform hypothesis test of mediation effect in causal mediation model with high-dimensional continuous mediators, such as "mediation", "BayesMed", and "hbayesdm".

The "mediation" package provides a function "mediate()" which can be used to perform hypothesis test of mediation effect in causal mediation model with high-dimensional continuous mediators. It allows to perform variable selection using Lasso, Ridge regression, or other methods.

The "BayesMed" package provides a function "bayes_mediation()" which can be used to perform Bayesian hypothesis test of mediation effect in causal mediation model with high-dimensional continuous mediators. It allows to incorporate prior knowledge and regularize the model parameters, which can help to identify the relevant mediators in a high-dimensional setting.

The "hbayesdm" package provides a function "hbayesdm()" which can be used to perform Bayesian hypothesis test of mediation effect in causal mediation model with high-dimensional continuous mediators.

1.11 Bayesian Mediation

Bayesian Mediation Analysis appeals to study for small samples. Due to this method, it is able to construct credible intervals for indirect effects for simple as well complex mediation models. It depends on two Bayesian sparse linear mixed models to analyze a large number of mediators simultaneously for constant exposure, and it is assumed that outcome being a small number of mediators is actively actual. This provides natural and more simple mediation analysis for multilevel models [18]. Hence it has many various vital advantages with the ability to incorporate prior information. On the other hand, it can help to improve the quality of estimates by integrating prior notification, which is essential with a moderate sample size.

1.12 High-Dimensional Mediation Models

Assume $T(x, m_1, m_2, ..., m_p)$ denote the survival time when exposure is x. Let

D_i: time from start to death event

C_i: potential censoring time

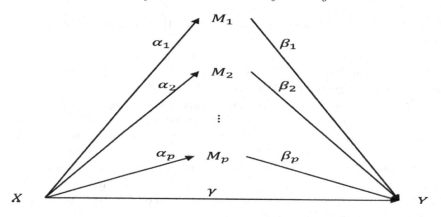

FIGURE 1.2: Graph describes high-dimensional mediation with the p mediators assumed to be uncorrelated with one another.

$T_i = min(D_i, C_i)$: observed survival time

$\delta_i = I(D_i \leq C_i)$: failure indicator where $I(.)$ is an indicator function

X_i: exposure

Z_i: other q baseline covariates

$M_i = (M_{1i}, M_{2i}, ..., M_{pi})^T$: a p-dimensional mediator vector for individual i where $i = 1, 2, ..., n$ and $p >> n$.

Figure 1.2 shows the relationship between the exposure, mediators, and the outcome event.

These models are primarily used to model the mechanisms of the exposure's effect on the outcome mediated by the mediators. With respect to time-to-event data, the rate at time t means the probability of experiencing death within the next unit of time, given that a patient is still alive before time t. Consider the regression model with high-dimensional mediators to denote the mediation effects.

$$\lambda_i(t|X_i, M_i) = \lambda_0(t)\exp[\gamma X_i + \theta^T Z_i + \beta_1 M_{1i} + ... + \beta p M_{pi}] \qquad (1.11)$$

where eq(10) is the Cox proportional hazard model which describes the relationship between the exposure X, mediators M, and time-to-event variable.

$$M_{ki} = c_k + \alpha_k X_i + \theta^T Z_i + e_{ki} \qquad (1.12)$$

where $k = 1, 2, ..., p$ and the above equation shows the influence of mediators by the exposure variables. Here,

$\lambda_0(t)$: baseline hazard function

Z: baseline covariates (e.g., age, gender, etc)

$\beta = (\beta_1, ..., \beta_p)^T$: the parameter vector relating the mediators to the outcome exposure effect.

$\alpha = (\alpha_1, ..., \alpha_p)^T$: the parameter vector relating the exposure to the mediators.

c_k: intercept term

$e_{ki} \sim N(0, \sigma^2)$: the residual.

1.13 Mediation Analysis Using R

In causal mediation analysis, the effect of an exposure (X) on the outcome (Y) is decomposed into three components: the direct effect (c), the indirect effect (ab), and the total effect (c + ab).

The direct effect (c) represents the effect of the exposure on the outcome that is not mediated by any other variable. It is the effect of X on Y when all the mediators (M) are set to their average value.

The indirect effect (ab) represents the effect of the exposure on the outcome that is mediated by one or more intermediate variables (M). It is the effect of X on Y through the mediator M.

The total effect (c + ab) represents the overall effect of the exposure on the outcome, including both direct and indirect effects.

In R, there are several packages that can be used to perform causal mediation analysis such as "mediation", "medflex", and "causMed". These packages provide functions for estimating the direct, indirect, and total effects as well as for testing the significance of these effects.

For example, the "mediation" package provides the "mediate()" function which can be used to estimate the direct, indirect, and total effects of an exposure on an outcome through a mediator. Additionally, the "mediate()" function provides confidence intervals and p-values for the effects.

It's important to note that for the causal inference it is important to have appropriate data and assumptions, and it's recommended to consult with experts in causal inference to ensure that the analysis is correctly performed.

> **Installation of packages**
>
> ```
> install.packages("devtools")
> devtools::install_github("dustinfife/flexplot")
> devtools::install_github("dustinfife/flexplot", ref="development")
> install.packages("mediation")
> library(mediation)
> ```

Installation of packages

```
set.seed(2014)
data("framing", package ="mediation")
med.fit= lm(emo ~ treat + age + educ + gender + income, data =
framing)
out.fit=glm(cong_mesg ~ emo + treat + age + educ + gender +
income,data = framing, family = binomial("probit"))
summary(med.fit)
summary(out.fit)
require(flexplot)
visualize(med.fit)
visualize(out.fit)
```

Compare results on mediation analysis

```
results=mediate(med.fit,out.fit,treat="treat",mediator="emo",
boot=TRUE,sim=500)
summary(results)
```

Nonparametric bootstrap confidence Intervals with the percentile method

Variable	Estimate	95% CI	P-value
ACME(contr)	0.08	[0.03,0.14]	0.000
ACME(treat)	0.08	[0.03,0.14]	0.000
ADE(contr)	0.01	[0.04,0.15]	0.81
ADE(treat)	0.01	[-0.09,0.16]	0.81
TotalEffect	0.09	[-0.10,0.24]	0.13
Prop.Medi(contr)	0.86	[-0.02,6.56]	0.13
Prop.Medi(treat)	0.88	[-7.16,6.03]	0.13
ACME(aver)	0.08	[-6.56,0.14]	0.000
ADE(aver)	0.01	[-0.09,0.15]	0.81
Prop.Medi(aver)	0.87	[-6.86,6.29]	0.13

Compare plots graphically

```
added.plot(emo ~ treat + age + educ + gender + income, data =
framing,method="lm")
mediate_plot(emo ~ age + educ + gender + income, data = framing)
```

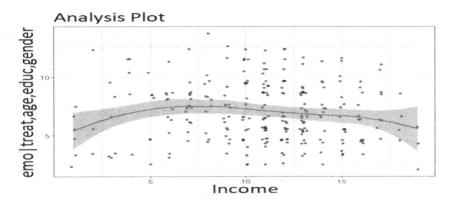

FIGURE 1.3: Fitplot description of model1.

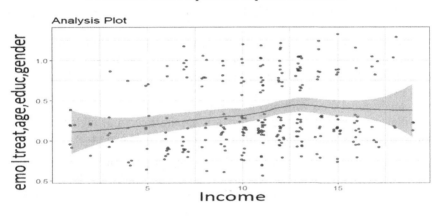

FIGURE 1.4: Fitplot description of model2.

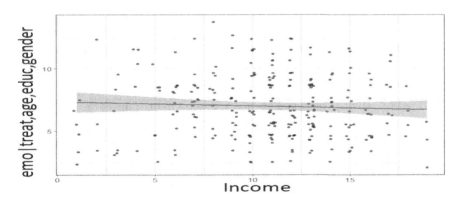

FIGURE 1.5: Comparing mediation plot graphically.

R packages include an R code, datasets, documentation, functions as well as other supporting files. We had successfully developed the R package named **"autohd: High Dimensional Bayesian Survival Mediation Analysis"** which can be downloaded through R studio and helped to study Mediation analysis. This R package is freely available through the Comprehensive R Archive Network (CRAN) at `https://CRAN.R-project.org/package=autohd` and runs on a variety of computing platforms. The first version of **autohd** package appeared at CRAN in 2021 with a significant number of new functionalities. This manuscript provides a detailed description of **autohd** package.To install **autohd** package, use the following standard R syntax

$R > install.packages(``autohd")$

where users may be prompted to select a CRAN mirror from which the package will be downloaded. This step needs to be done only once (unless one wishes to update the autohd package to the new version).

Explicitly mediation analysis provides time to event data analysis using methods such as Cox proportional hazard model and accelerated failure time model to study high-dimensional data using Bayesian inference. To compute mediation analysis, we used the data based on head and neck cancer data which is described as High-Dimensional gene expression data. Various functions were mapped with this dataset. Primarily we incorporated function with high-dimensional data with accelerated failure time using mediation analysis, high-dimensional data with missing data imputations, and high-dimensional competing risk using mediation analysis with different aspects.

1. hnscc: High-Dimensional head and neck cancer gene expression data. This data involved 565 patients, including 104 variables. It includes patients' ID, left censoring, death, survival event with competing risk, overall survival, duration of progression-free survival, progression event, and high-dimensional covariates GJB1,..., HMGCS2.

2. hnscc2: High-Dimensional head and neck cancer gene expression data. This data involved 565 patients, including 104 variables. It includes patients' ID, left censoring, death, survival event with competing risk, overall survival, duration of progression-free survival, progression event, and high-dimensional covariates GJB1,..., HMGCS2.

3. srdata: High-Dimensional gene expression data. This data included 288 rows and 252 variables. It comprises the ID of subjects involved, the number of observations recorded, death event one if died and death event 0 if alive, duration of overall survival, information based on left censoring, and high-dimensional covariates C6kine,..., GFRalpha4.

Various functions are avaliable in the autohd package, which gives us convenient results using the mediation effect. For example, **"coxmulti"** provides the easy way to work with multivariate Cox proportional hazard model with high-dimensional data analysis. It supports obtaining the relation between the event incidence, as expressed by the hazard function and a set of covariates.

It works to relate the several risk factors associated with survival time. Mediation analysis indicates whether the effects of X (the independent variable) on Y (the dependent variable) operate through a third variable, M (the mediator). The function "coxmulti" gives the hazard rate, UCL and LCL, and P-value of the five covariates for the event death.

Multivariate accelerated failure time data appears when the subject under study experience several types of failures or recurrences of a specific phenomenon. This function **"hdraftma"** is useful to work with high-dimensional data where the effect of covariates is used to accelerate or decelerate the course of the disease by some constant utilizing the approach of Bayesian analysis. Similarly, it works to fit the AFT model if the variables' dimension and survival information are accessible. It filters the significant variables and performs mediation analysis with filtered-out variables.

Cox Proportional model (Cox, 1972) is the most commonly used multivariate approach for analyzing survival time data in medical research. This method works for both quantitative predictor variables and categorical variables. Multivariate Cox Proportional Model describes the relationship between the event incidence, as expressed by the hazard function and a set of covariates. This function **"hdcoxma"** clears the significant variables by fitting multivariate cox proportional model, which involves five variables within those that provide a handful of variables with their alpha. Values and beta values are mediator model exposure coefficients. High-dimensional survival analysis using SurvM-Cmulti: The primary survival analysis approach in the presence of censored data indicates that event had not occurred during the study. High-dimensional data is when the number of variables exceeds the number of observations.This function **"hdsurvma"** filters the significant variables with high number of iterations. Later it performs mediation analysis with the significant variables and provides a handful with their alpha. Values are mediator model exposure coefficients and beta.

High-dimensional univariate cox proportional hazard analysis. Univariate Cox Proportional hazard involves single-variable relation of predictor variables. However, single-variable relations are very informative due to relations among the values of the predictors and potential interactions among the predictors concerning outcome. The function **"hidimcox"** performs Cox PH using given variables and survival information 1.2.

R package "missForest" is used in mixed-type data for imputing missing values in continuous and categorical data, which includes complex interactions and non-linear relation if survival information and dimension of variables is available; following functions perform the process of replacing missing data with substituted values using missForest function and filters significant variables. They are as follows:

imphdaft: AFT model with Weibull distribution
imphdcox: Multivariate CoxPH model with 5 variables
imphdsurv: Univariate survival analysis with higher number of iterations.
impuni: Univariate survival analysis with higher number of iterations.

TABLE 1.1:   Selected variables on different "autohd" functions

Function	Selected variables	beta	alpha
imphdaft	Adiponectin	0.02	0.99
imphdsurv	Adiponectin	0.99	0.78
	AgRP	−4.26	−0.20
	ALCAM	−0.01	0.06
	Angiogenin	−0.01	0.24
	Angiopoietin1	−0.09	−0.20
hdaftma	KIFIA	−0.56	1.01
hdraftma	AFF3	0.98	0.37
	C3orf15	−0.74	1.00
hdcoxma	AFF3	−0.84	0.97
hdsurvma	TOX3	0.97	0.20

TABLE 1.2:   Hazard rate with CI and P-value of the selected variables

Function	Variables	HR	LCL	UCL	p-value
multicox	PGC	1.13	0.10	1.27	0.050
	C7	1.01	0.97	1.06	0.860
	HPN	0.99	0.94	1.06	0.860
	DDC	1.05	0.97	1.14	0.190
hidimcox	GJB1	1.02	0.96	1.08	0.593
	PPP1R9A	1.01	0.96	1.06	0.748
	HPN	1.02	0.96	1.08	0.529
	SLC4A4	1.02	0.96	1.09	0.450
	HNF1B	1.07	0.93	1.24	0.340

impuniaft: Univariate AFT

impunicox: Univariate CoxPH model

These functions further perform mediation analysis within the significant variables and provide a handful of variables with their alpha. Values are the mediator model exposure coefficients and beta coefficients.

Competing risk occurs in the analysis of survival data. A competing risk is an event whose occurrence prevents the occurrence of the primary event of interest. It considers event times due to multiple causes or more than one event type. Variable selection is a difficult task and even more in the presence of competing risk with high-dimensional data in survival analysis.

unicrma: Univariate AFT model with Weibull distribution

unihdaftma: Univariate AFT model with mediation analysis

unihdcoxma: Univariate Cox PH model

unihdma: perform Survival analysis with higher number of iterations.

The output of the above functions regarding competing risk analysis shows no active variables.

In medical research, unique challenges are seen due to a large number of information of patients. There is frequent use of mediation analysis for the time-to-event outcome to examine possible mechanisms. This kind of analysis will help researchers explore many casual paths instead of simple informal ways. Zhang introduced two procedures for mediator selection with high-dimensional exposures and high-dimensional mediators [19]. Mediation is one of the techniques in which researchers can explain the process by which one variable affects the other. Primarily mediating variables are popular in psychology due to their historical dominance of the stimulus organism response model [20]. Similarly, it is widely applicable in Epidemiology. For example, mediation analysis can be used to investigate BMI as a mediator of the relation between smoking and insulin levels [21], or to analyze food expenditures as a mediator of the relationship between socio-economic status and healthiness of food choices.

In this chapter, various functionalities of **autohd** package, which helped applied researchers to conduct mediation analysis for high-dimensional data are presented. The package implements mediation analysis for high-dimensional data and is a statistical tool in the Bayesian study. Here Cox PH model, accelerated failure time model using Bayesian inference. Survival data analysis methods such as the Cox proportional hazard model and the accelerated failure time model using the Bayesian approach are studied. Missing data imputation techniques tool to work with high-dimensional data coupled for mediation analysis presented by the active mediator variables. We hope this **autohd** package can distribute as a helping hand for other researchers to add more to high-dimensional data in survival analysis. There are specific questions for high-dimensional which are to be answered and are choices for future studies.

Chapter 2

Bayesian Mediation Analysis

2.1 Introduction

Methods proposed on mediation analysis are mostly based on the frequentist perspective [22, 23, 24, 25, 10, 26]. Recently, some extension has been carried out with Bayesian inference by computing the posterior distribution of the products of coefficient and preparing inference on the causal effects of interest [27, 28, 29, 30, 31]. The non-parametric bounds of principal strata on the dichotomous mediator and dichotomous outcomes are prepared by the prior distribution of possible range [28]. The mediator distribution is attempted by the Dirichlet process of the mixture model. Mostly, the causal effects are attempted by principal causal effects, natural direct and indirect effects. However, the Bayesian non-parametric method has not been explored for the natural direct and indirect impact.

The reason to use Bayesian mediation analysis is the availability of several features to work within this direction. It works nicely in the multilevel mediation analysis, which is not difficult to work with the conventional frequentist approach. Now the Bayesian approach is not an attractive choice in multilevel modeling. It is also attractive conceptually and becomes powerful with the Markov Chain Monte Carlo (MCMC) approach. Software with OpenBUGS and R strengthen it with computational advancement. It becomes an attractive choice because it is a free large sample approximation. The inference is strong enough for small sample data. The features available with Bayesian make it appealing for a small sample size. Sometimes simulation studies prove that Bayesian mediation is more potent than the frequentist approach. It shows that the Bayesian noninformative prior provides similar outcomes as the frequentist method. Bayesian works nicely for testing sensitivity toward assumed the prior value by changing prior distribution values. The Bayesian often helps to accommodate the prior information through mediated effects from different studies. It may provide different results by changing the power of the study. The objective of the consideration of different prior densities helps us to cover different noninformative and conjugate priors by model parameters in the single-level and multilevel mediation model. Other types of prior distribution stand with Cauchy distribution [32] as an alternative to the normal distribution to generate posterior estimates of the regression

DOI: 10.1201/9781003298373-2

20

coefficients. However, it is anticipated that the likelihood will make an impact on the Bayesian inference [33] than prior. So assumptions with priors become irrelevant while probability takes the leading role in the inference. If the prior is incorrect, the Bayesian mediation estimate becomes biased. So prior selection technique requires scrutinizing and making it appropriate. The standard approach while quantifying prior information is based on understanding the associated uncertainty of the preceding information. The strong prior that influences the likelihood should be avoided. Inference requires to be taken from observed data. Sometimes the dispersed priors can be considered as an alternative of the Bayesian inference from bias. Sometimes, the sensitivity can also be considered to explore the influence of different prior assumptions. It is suggested to have noninformative prior and combine the information from the meta analysis [34]. The consideration of previous evidence from conducted studies also useful to form correct hypothesis and suitable prior assumption.

Mediation analysis is a statistical technique used to investigate intermediate variables that explain the relationship between an independent variable and a dependent variable. However, interpreting the causal effects of mediation analyses can be challenging due to the possibility of unmeasured confounders of the mediator-outcome relationship, even with randomized subjects to the independent variable. Additionally, commonly used frequentist methods for mediation analysis calculate the probability of the data given the null hypothesis, rather than the probability of a hypothesis given the data, as in Bayesian analysis.

To address these issues, the potential outcomes framework can be applied to mediation analysis under certain assumptions to compute causal effects. Bayesian mediation analysis can also provide probabilistic interpretations of indirect effects.

The steps in Bayesian causal mediation analysis are demonstrated through an empirical example. This approach improves the interpretation of causal effects in mediation analysis and provides a more comprehensive understanding of the relationships among variables. Recently the causal mediation analysis received more attention in different disciplines [35, 36]. It shows the effect of the intermediate variable on the exposure variable to explore the results of the outcome. Mediation helps to measure the impact by direct and indirect effects. It builds on rigorous assumptions of the exposure-outcome and exposure-mediator. Similarly, it explores the mediator-outcome relationship that supports the usage of the classical formulas in the regression setting [35, 37], by exploring the relationship between the exposure-mediator, exposure-outcome, and mediator-outcome [1, 38]. There are several frameworks to explore the causal inference for continuous [25] and dichotomous variables [36]. Recently, there has been an attempt toward exploring several mediator variables, but standard practice is to work with two or three mediator variables [39]. Sometimes, several ad hoc approaches work to fit the univariate mediation models [40, 41] or by summarizing the effects across different mediators. This approach avoids correlation and estimates that the mediation effect does not have a causal interpretation or is restricted by the limited number of mediator variables.

2.2   Mediation Analysis Example

Suppose the mediating effect is presented for depression, and it is linked to disease status and quality of life among cancer patients. One hypothesis is presented as better quality of life reduced depression before measuring the outcome of interest. Now the quality of life can be related with treatment provided with two potential value i.e. $M_i(1)$ and $M_i(0)$, only one of which will be observed, that is, $M_i = M_i(T_i)$. For illustration, if the patient is treated, then $(T_i = 1$ and it is observed as $M_i(1)$ but not $M_i(0)$. Now we can define the potential outcomes as it depends on the mediator and the treatment variable. It presented as $Y_i(t, m)$ by denoting the potential outcome that would result if the treatment and mediating variables equal t and m, respectively. Now the presentation $Y_i(1, 1.5)$ as the degree of depressive symptoms that would be observed in the cancer patients i can perform during the treatment of the self-efficacy score of 1.5. Now the observed outcome Y_i equals to $Y_i(T_i, M_i(T_i))$. Finally, if there is no interference between units is assumed throughout. The potential mediator values of each unit do not depend on the treatment status of the other units. We can define the causal mediation effects for each unit i as follows.

$$\delta_i(t) = Y_i(t, M_i(1)) - Y_i(t, M_i(0)) \tag{2.1}$$

for $t = 0, 1$. Now the causal mediation represents the indirect effect of the treatment on the outcome by the mediating variable [42, 37, 43]. This one helps to understand that the one unit change of the mediator value would be realized under the condition $M_i(0)$ toward evaluating the observed treatment condition $M_i(1)$. It holds the treatment status at t? If the treatment has no effect on the mediator then it is represented as $M_i(1) = M_i(0)$, while the causal mediation effect becomes zero. Now the term $Y_i(t, M_i(t))$ is observable for units with $T_i = t, Y_i(t, M_i(1 - t))$ that can never being observed for any unit.

2.3   Causal Mediation Analysis

Conventionally the causal mediation analysis is presented, formulated with linear structural equation modeling [1, 44, 45, 46, 38, 47]. However, it is challenging due to a framework that cannot be generalized because causal mediation affects the application beyond the specific statistical models. It states the key identification assumption in the context of a particular model, and quite difficult to separate this limitation at the research design formulation stage. Secondly, the method prepared cannot be generalized in the non-linear context with logit and probit models with discrete mediator and outcome with non- or semiparametric models.

The causal mediation analysis is presented by the single exposure of interest with high-dimensional setting, where candidate mediators mediate with the effect of the exposure on the outcome variable. Suppose our study is based on the $i, i = 1, ..., n$, subjects toward collecting the exposure information A_i, p. Candidate mediators $M_i = (M_i^{(1)}, M_i^{(2)}, ..., M_i^{(p)})^T$ are presented as outcome variables Y_i with q covariates as $C_i = (C_i^{(1)}, ..., C_i^{(q)})^T$. We can look into the outcome variable Y_i and mediator M_i as continuous variable. The framework is presented along with mediators and their causal effects. Suppose $M_i^j)(a)$ define the potential value of the jth mediator as $j = 1, ..., p$, for the ith subject. The exposure variable is presented as K levels with $K \times p$ potential counterfactual random variables and it is defined as $M(1)(1), M(2)(1), ..., M(p)$ as $Y_i(a, m) = Y_i(a, m^{(1)}, ..., m^{(p)})$. Sometimes, a total of n subjects $i, i = 1, ..., n$, linked with the exposure are a with the mediators as $m = (m^{(1)}, ..., m^{(p)})$. The joint effect of the mediator is presented as $M_i(a) = (M_i^{(1)}(a), M_i^{(2)}(a), ..., M_i^{(p)}(a))$. However, it is difficult to observe the real data along with the hypothesis. In this framework, the Stable Unit Treatment Value Assumption (SUTVA) assumption in the causal inference makes suitable charecteristic for the observed treatment level as $M_i = \sum_a M_i(a)I(A_i = a)$, and $Y_i = \sum_a \sum_m Y_i(a, m)I(A_i = a, M_i = m)$, where $I(.)$ is the indicator function. We can also present it as $Y_i(a) = Y_i(a, M_i(a))$. Further, the total effect, natural direct and indirect effects can be formulated as

$$Y_i(a) - Y_i(a^*) = Y_i(a, M_i(a)) - Y_i(a^*, M_i(a^*)) \tag{2.2}$$

$$Y_i(a, M_i(a)) - Y_i(a^*, M_i(a^*)) = Y_i(a, M_i(a)) - Y_i(a, M_i(a^*)) + $$
$$Y_i(a, M_i(a^*)) - Y_i(a^*, M_i(a^*)) \tag{2.3}$$

$$Y_i(a, M_i(a)) - Y_i(a, M_i(a^*)) + Y_i(a, M_i(a^*)) - Y_i(a^*, M_i(a^*)) = \text{NIE} + \text{NE} \tag{2.4}$$

This causal effects are formally defined in terms of potential variables, which are not necessarily observed, but the identification of causal effects must be based on observed data.

2.4   Bayes Theorem

Bayes' theorem is a mathematical formula used to calculate the probability of an event based on prior knowledge of conditions that might be related to the event. It is named after English statistician Thomas Bayes (1701–1761). Bayes' theorem provides a way to revise existing probabilities when new evidence is acquired. It follows a logical structure known as Bayesian inference.

The theorem is stated mathematically as:

P(A—B) = P(B—A) * P(A) / P(B)

Where: P(A—B) is the conditional probability of event A given that B is true, P(B—A) is the conditional probability of event B given that A is true, P(A) is the prior probability of event A, and P(B) is the prior probability of event B.

2.4.1 Bayes' Rule

The probability statement of θ given y can be presented by probability density function and can be written as a product of two densities that are often defined as prior distribution of $p(\theta)$. Joint probability density or mass function is prepared of two densities as the prior distribution $p(\theta)$ and the sampling distribution of $p(y|\theta)$ respectively as

$$p(\theta, y) = p(\theta)p(y|\theta) \tag{2.5}$$

The sampling condition on the known value of the data y using the basic property of conditonal probability with Bayes' rule presented with posterior density as

$$p(\theta|y) = \frac{p(\theta, y)}{p(y)} = \frac{p(\theta)p(y|\theta}{p(y)} \tag{2.6}$$

where $p(y) = \sum_\theta p(\theta)p(y|\theta)$, and as the sum is over all possible values of θ (or θ $p(y) = \int p(\theta)p(y|\theta)d\theta$ in the case of continuous θ). An equivalent form of (1.1) omits the factor $p(y)$, which does not depend on θ and, with fixed y, can thus be considered a constant, yielding the unnormalized posterior density by

$$p(\theta|y) \propto p(\theta)p(y|\theta) \tag{2.7}$$

Now the simple expression defined the core of Bayesian inference by the primary task of any specific application toward developing the model $p(\theta, y)$ and perform the necessary computations to summarize $p(\theta|y)$ inappropriate ways.

2.4.2 Prediction

Bayesian prediction is the process of using Bayesian inference to make predictions about future events. Bayesian inference is a type of statistical inference that uses Bayesian probability theory to make predictions. It is based on the idea that the probability of an event occurring is based on both prior knowledge and new evidence. Bayesian inference can be used to make predictions about unknown data by combining prior beliefs about the data with new information. We follow a similar logic to make inferences about an unknown observable, often called predictive inferences. Before the data y are considered, the distribution of the novel but observable y is

$$p(y) = \int p(y, \theta)d\theta = \int p(\theta)p(y|\theta)d\theta \qquad (2.8)$$

It is the marginal distribution of y and is more informative and named as the prior predictive distribution. The prior is conditional on a previous observation of the process. It is predictive because the distribution for a quantity is observable. Once the data y is observable after the data y have been observed then it can predict as an unknown observable $\hat{y}$ from the similar process. Now $y = (y_1, ..., y_n)$ becomes the vector of the recorded weights of an object weighed n times on a scale as $\theta = (\mu, \sigma^2)$. It is the unknown accurate weight of the object and the measurement variance of the scale, and $\hat{y}$ can be recorded as the weight of the thing in a planned new weighing. Now the distribution of $\hat{y}$ is defined as the posterior predictive distribution. It is posterior because it is conditional on the observed y and the predictive as it is a prediction for an observable $\hat{y}$.

$$p(\hat{y}|y) = \int p(\hat{y}, \theta|y)d\theta \qquad (2.9)$$

$$p(\hat{y}|y) = \int p(\hat{y}, \theta, y)p(\theta|y)d\theta \qquad (2.10)$$

$$p(\hat{y}|y) = \int p(\hat{y}|\theta)p(\theta|y)d\theta \qquad (2.11)$$

The second and third lines display the posterior predictive distribution as an average of conditional predictions over the posterior distribution of θ. The last equation follows because y and $\hat{y}$ are conditionally independent given θ in this model.

2.5   Likelihood

Bayesian likelihood is a probability distribution used to estimate the likelihood of a given event. It is based on the Bayes theorem, which states that the probability of an event is equal to the probability of the event given the data, multiplied by the prior probability of the event. The Bayesian likelihood is calculated by taking the product of the prior probability of the event and the likelihood of the data given the event. This is used to estimate the probability of the event occurring based on the data collected. The Bayes' rule defined with probability model means as y affect the posterior inference only through the function $p(y|\theta)$, which, when regarded as a function of θ, for fixed y, is called the likelihood function. In this way Bayesian inference obeys what is sometimes called the likelihood principle, which states that for a given sample of data, any two probability models $p(y|\theta)$ that have the same likelihood function yield the same inference for θ.

2.6   Likelihood and Odds Ratios

The ratio through the posterior density of $p(\theta|y)$ examined at the points θ_1 and θ_2. It is under a given model is presented by the posterior odds of θ_1 as compared to θ_2. Now the familiar application of the discrete parameter is defined as θ_2 as taken from the complement of θ_1.

Odds ratios are a measure of the strength of the association between two variables in a categorical data set. They are calculated by taking the ratio of the odds of an event occurring in one group compared to the odds of the event occurring in another group. For example, if the odds of a person developing a certain disease is 1:5 in group A, and 1:10 in group B, the odds ratio would be 1.5 (5/10).

Now the odds gives an alternative of the probabilities and have the attractive choice of the Bayes's rule by

$$\frac{p(\theta_1|y)}{p(\theta_2|y)} = \frac{p(\theta_1)p(y|\theta_1)/p(y)}{p(\theta_2)p(y|\theta_2)p(y)} = \frac{p(\theta_1 p(y|\theta_1)}{p(\theta_2 p(y|\theta_2)} \tag{2.12}$$

The posterior odds is obtained by the prior odds multiple through the likelihood ratio $p(y|\theta_1)/p(y|\theta_2)$. Likelihood is a measure of how likely it is that a particular event or outcome will occur, while odds ratio is a measure of the probability of an event occurring compared to the probability of it not occurring. The odds ratio is calculated by dividing the probability of an event occurring by the probability of it not occurring. For example, if the probability of an event occurring is 75%, then the odds ratio would be 3:1, since the probability of it not occurring is 25%.

2.7   Mediation Analysis in Clinical Research

In clinical research, mediation analysis is used to understand the mechanisms through which an intervention or treatment affects an outcome. There are several different methods for performing mediation analysis in clinical research, which can be broadly classified into four categories:

Path analysis: This method is based on structural equation modeling (SEM) and it models the relationships between the exposure, the mediator, and the outcome as a system of equations. Path analysis can be used to estimate the total, direct, and indirect effects of the exposure on the outcome, and it can also be used to test hypotheses about the relationships between the variables.

Natural experiments: This method uses natural variation in the exposure or the mediator to estimate the causal effects of the exposure on the outcome. Natural experiments can be used to estimate the total, direct, and indirect effects of the exposure on the outcome, and they can also be used to test hypotheses about the relationships between the variables.

Instrumental variable (IV) analysis: This method uses an instrumental variable (IV) as a proxy for the exposure to estimate the causal effects of the exposure on the outcome. IV analysis can be used to estimate the total, direct, and indirect effects of the exposure on the outcome, and it can also be used to test hypotheses about the relationships between the variables.

Causal inference: This method uses causal inference techniques, such as propensity score matching or inverse probability weighting, to estimate the causal effects of the exposure on the outcome.

2.8   Bayesian Mediation Analysis Using R

Bayesian mediation analysis is a method for estimating the direct and indirect effects of an exposure on an outcome, using a Bayesian framework. This approach allows for the incorporation of prior information and can provide more robust estimates, especially when the sample size is small or the data is noisy.

In R, there are several packages that can be used to perform Bayesian mediation analysis, such as "bayesMed", "MCMCmediation", and "mediation".

The "bayesMed" package provides a function "bayes_mediation()" which can be used to perform Bayesian mediation analysis. The function estimates the direct and indirect effects using a Bayesian framework and provides summary statistics such as the posterior mean, standard deviation, and credible intervals for the effects.

The "MCMCmediation" package provides functions "mediation_mcmc" and "mediation_ab" which can be used to perform Bayesian mediation analysis using Markov Chain Monte Carlo (MCMC) methods. These functions provide posterior samples of the direct and indirect effects, which can be used to estimate summary statistics such as the mean and credible intervals.

The "mediation" package also provides a function "mediation_bayes()" which can be used to perform Bayesian mediation analysis. This function estimates the direct, indirect, and total effects using a Bayesian framework and provides summary statistics such as posterior mean, standard deviation, and credible intervals for the effects.

It's important to note that Bayesian mediation analysis requires specifying a prior distribution for the parameters and it's recommended to consult with experts in Bayesian statistics to ensure that the analysis is correctly performed.

2.9 Mediation in Epigenetic Using R

In epigenetic studies, researchers often investigate the role of specific genes in mediating the effect of an exposure on a phenotype. Gene-based mediation analysis is a method that can be used to identify genes that mediate the effect of an exposure on a phenotype by testing the association between the exposure, the gene, and the phenotype.

One approach for performing gene-based mediation analysis is to use a gene-based test, such as a gene set enrichment analysis (GSEA) , which tests whether a gene set (e.g., all genes in a pathway) is associated with the exposure, the phenotype, or both. Another approach is to use a single-gene test, such as a linear regression or a logistic regression, to test the association between the exposure, the gene, and the phenotype for each gene individually.

In R, there are several packages that can be used to perform gene-based mediation analysis in epigenetic studies, such as "mediation", "Mediation-Tools", and "genomed".

The "mediation" package provides a function "mediate()" which can be used to perform gene-based mediation analysis. It can be used to test the association between the exposure, the gene, and the phenotype using linear or logistic regression models, and it allows to adjust for confounding variables.

The "MediationTools" package provides functions for performing gene-based mediation analysis. It can be used to test the association between the exposure, the gene, and the phenotype using linear or logistic regression models, and it allows to adjust for confounding variables.

The "genomed" package provides functions for performing gene-based mediation analysis using GSEA. It can be used to test the association between the exposure, the gene set, and the phenotype.

Chapter 3

Parametric Survival Analysis

3.1 Introduction

Parametric survival analysis is a statistical method used to analyze time-to-event data, where the underlying distribution of the time-to-event variable is assumed to follow a specific parametric distribution such as exponential, Weibull, or log-normal distributions. In these models, the parameters of the distribution are estimated using maximum likelihood estimation, and the probability of an event occurring is modeled using the cumulative distribution function (CDF) of the assumed distribution.

The most commonly used parametric survival models are the exponential model, Weibull model, and the log-normal model.

The exponential model assumes that the time-to-event variable follows an exponential distribution and is characterized by a single parameter, the hazard rate, which represents the instantaneous risk of an event occurring.

The Weibull model assumes that the time-to-event variable follows a Weibull distribution and is characterized by two parameters, the shape parameter and the scale parameter. This model is useful for modeling right-skewed data, where the hazard rate increases or decreases over time.

The log-normal model assumes that the logarithm of the time-to-event variable follows a normal distribution and is characterized by two parameters, the mean and the standard deviation.

Parametric survival analysis can be performed using various software such as SAS, R, or STATA, and it allows for the estimation of survival probabilities, hazard ratios, and median survival times. This method is useful when the underlying distribution of the time-to-event variable is known or assumed, and there is a good fit between the data and the assumed distribution. Bayesian approaches had gained a wide popularity in medical, pharmaceutical, and social science research because it allows researchers to combine prior information with data to model data processes to incorporate prior knowledge of likelihood of an event to interpret the trail results [48]. Survival data analysis resembles to life-time data which takes into account time of the events in various fields. In practical situations, often the data is censored. Cox's proportional hazard(PH) [49] model is one of the most popular modeling techniques in survival analysis. The Cox PH model is similar to a multiple linear regression that explores the

DOI: 10.1201/9781003298373-3

relationship between a hazard and related independent explanatory variables over a period of time. It describes the impact of a risk factor on a treatment of patient through a parameter called hazard ratio(HR). High-dimensional data is defined as set of data points where the number of features p are larger than the number of observations N as $p >> N$ [9]. Such type of data have extensive number of variables with comparatively few subjects with the objective to inspect different patterns within the datasets. This type of data is very common in healthcare industry where the number of features are exclusively more in number (i.e., blood sugar, height, weight, etc). Visual exploration is an attractive way to explore the conditional dependence structure between the different variables for high dimensional data and establish the possible relation by other linear models. There are two choices to study handle high-dimensional data by (I) features selection procedure and (II) regularization procedure. In case of features selection procedure, it only includes few features from the dataset. There are many ways to drop features from the datasets like drop features with many missing values, with low variance and with low correlation with the response variable [50]. The regularization method is an alternative to deal with high-dimensional data by using regularization method i.e., without dropping the features. It works through—Principal Components Analysis, Principal Components Regression, Ridge Regression, and Lasso Regression. Each of these techniques can be effectively used to deal with high-dimensional data.

Censoring is a prevalent condition in survival analysis when information of time to outcome event is not known for all the study participants. This is applicable to missing data where time to event is not observed due to various reasons like study is terminated since subjects under study had left before experiencing an event. There are three general types of censoring— Right censoring, left censoring, and interval censoring. A survival data is said to be **Right-censored** if the survival time of the individual goes beyond some particular value. This happens commonly when the subjects involved in the study had not experienced the event of interest. For instance, consider the variable of interest as survival time i.e., the endpoint of subject is death then the survival data can be right-censored if some of the subjects are alive by the end of study while some of them died during the study follow-up [51]. Alternatively, **left censoring** is seen when actual survival time of a subject is less that the observed survival time. For example, if few children had assigned to a specific task in order to check the one's ability, at the time of selection some of them might know how to do that task and thus from the birth time to execution of the task is left-censored for these children [52]. Our study mainly focused on another censoring called **Interval censoring**. Interval censoring is seen when variable of interest i.e., time is not observed directly and at a specific time interval it is known to occur [53]. One of the particular case of interval censoring is Right censoring. Suppose T be the survival time of study and it is interval censored then interval censoring will be denoted as $I = (L, R]$.

3.2 Bayesian Cox Proportional Hazard Model in Survival Analysis

Assume a sample of n individual for i^{th} time possible censored survival times be $Y_1 \leq Y_2 \leq Y_3 \leq ... \leq Y_n$. If Y_i is an observed death then $\delta_i = 1$ and $\delta_i = 0$ if it was a right-censored event. This means that at Y_i time individual was alive but seen at that time. Let true survival or failure times be $T_i (1 \leq i \leq n)$ then

$Y_i = T_i$ if $\delta_i = 1$ and $Y_i \leq T_i$ if $\delta_i = 0$

Here T_i true failure time which is unknown such as

$Y_i = T_i$ if $\delta_i = 1$ and

$Y_i \leq T_i$ if $\delta_i = 0$

Suppose that d-dimensional covariate vectors $X_1, X_2, ..., X_n$ for n individuals for possible censored survival times. To estimate the covariate effect X_i on the true survival times T_i, let number of deaths and censored events be denoted as d_j and a_j, respectively. Let lifetime of a random individual with covariate X be Y. Then survival function will be given by

$$S_X(t) = P_x(Y \geq t) = exp(-H_X(t)) \qquad (3.1)$$

$$S_X(t) = exp^- \int_0^t h_X(d_y) \qquad (3.2)$$

In above equation, $H_X(t)$ one form of cumulative hazard right continuous function with $H_Z(t) = 0$ and $h_X(d_y)$ be the instantaneous hazard measure. The objective of this model is to establish β with an aim to test each component of β so that we can find if that component of X is stochastically significant effect on the survival times. The likelihood function for the data is given by

$$L = (\prod_{\delta_i=0} P_{Xi}(Y > Y_i))(\prod_{\delta_i=1} P_{X_i}(Y = Y_i)) \qquad (3.3)$$

In order to simplify the likelihood function suppose $\Delta j > 0$ and $\widetilde{Y_j} + \Delta j < \widetilde{Y}_{j+1} + \Delta j + 1$ for all j. Therefore likelihood will be

$$L_\Delta = \prod_{i=1}^n \begin{cases} P_{Zi}(Y > \widetilde{Y_j} + \Delta_j) & if \delta_i = 0 \\ P_{Zi}(\widetilde{Y_j} - \Delta_j) < Y \leq \widetilde{Y_j} + \Delta_j) & if \delta_i = 1 \end{cases} \qquad (3.4)$$

By definition, the true lifetime $T_i > \widetilde{Y_j}$ for censored individuals with $Y_i = \widetilde{Y_j}$ so that equation (4) is the appropriate probability if the $\Delta_j > 0$ are sufficiently small. The above likelihood equation (4) can be mutually proportional to equation (3) with the limit $\Delta_j \to 0$. The above equation can be written in terms of survival function $S_X(t)$ as

$$S_X(t) = \prod_{i=1}^{n} \begin{cases} S_{X_i}(Y_{f_j} + \Delta_j) & if \delta_i = 0 \\ S_{X_i}(Y_{f_j} - \Delta_j) - S_{X_i}(Y_{f_j} + \Delta_j) & if \delta_i = 1 \end{cases} \qquad (3.5)$$

3.3   Markov Chain Monte Carlo(MCMC) Method

Markov Chain Monte Carlo algorithms were originally developed in the 1940s by physicists at Los Alamos [54]. Over the past few years, MCMC methods had reorganized statistical computing. These methods had affected the study of Bayesian approach completely by permitting the complex models to be used in impressive way. Explicitly not only Bayesian are one to be convenient for studying MCMC but there is extending in other statistical situations [55]. Markov Chain Monte Carlo is a simulation method for sampling from Posterior distributions, which computes posterior quantities of interest. This method is used in Bayesian assumption to obtain information based on distribution to estimate posterior distribution.

In Bayesian statistics, the recent development of MCMC methods has made it possible to compute large hierarchical models that require integrations over hundreds to thousands of unknown parameters [56]. These methods create samples from a continuous random variable with probability density proportional to a known function. This method provides an alternative approach to random sampling, a high-dimensional probability distribution where the next sample is dependent upon the current sample. This method is highly intractable to high-dimensional distributions. The main objective of this method is to initiate the samples that are proportional to posterior distribution in order to come up with optimum estimate. The basic idea behind this method is: Given the probability distribution π on a set Ω with distribution π. This method does by constructing a Markov Chain with stationary distribution π and stimulating the chain [57].

Definition: Countable (finite) state space Ω. Sequence of random variables X_t on Ω for t=0,1,2,... . X_t is a Markov Chain if

$$P[X_{t+1} = y | X_t = x_t, ..., X_0 = x_0] = P[X_{t+1} = y | X_t = x_t] \qquad (3.6)$$

Notationally, $P[X_{t+1} = i | X_{/t} = j] = p_{ij}$ where time is homogeneous. MCMC methods are much helpful Monte Carlo method which are usually applied in Bayesian approach. There are two main prominent algorithms for MCMC methods: (i) the Metropolis-Hastings algorithm and (ii) the Gibbs sampling algorithm.These methods does not work properly if there are multi modes in relative probability distribution.

3.4 Credible Interval

Bayesian statistics are constructed according to the idea that probability of happening is affected by prior expectation of probability. In practical situations, it is important to report confidence interval for reporting the study evidence. Confidence Interval containing upper and lower limit values shows the variability around the estimates of effects. Considering the Bayesian approach an unobserved value of the parameter that falls within a specific probability is defined as Credible interval. The Bayesian credible interval (CrI) is quite comparable to the Confidence interval (CI) in the frequentist approach [58]. This approach means that as the number of trials increase the relative frequency will drop off. In order to construct credible interval, we need to truncate left tail or right tail or both probability of a posterior distribution so that remaining probability mass function is obtained. A Bayesian credible interval of size $1 - \alpha$ in an interval (a, b) such that

$$P(a \leq \theta \leq b|x) = 1 - \alpha.p(\theta|x)d\theta = 1 - \alpha \qquad (3.7)$$

Thus, $100(1-\alpha)\%$ credible interval C defines the subset of parameter space called as θ such that the integral is given by

$$\int_C \pi(\theta|X)d\theta = 1 - \alpha \qquad (3.8)$$

This interval is from the domain where the probability distribution of a random variable which is on the conditional on the fact that it is obtained from the study [59]. Bayesian credible intervals are classified in two categories—equal-tail interval and highest posterior interval. **Difference:** The Bayesian credible interval is comparatively more practical with the confidence interval. Confidence interval which represents variability of the interval can be modified as probabilistic statement with respect to true values [60]. Credible interval denotes the change in the location parameter and hence probabilistic statement can be estimated clearly.

3.5 High-Dimensional Bayesian Variable

There are various methods recently introduced to deal with high-dimensional data using Bayesian approach. Some of them are Gibbs Variable Selection (GVS), Stochastic Search Variable Selection (SSVS), Laplacian prior, reversible jump using MCMC, etc [61]. Rather than selecting a model, the recent studies had shifted it to the form of parameter estimation in

Bayesian methodology. It tries to estimate the posterior probability i.e., the effort is made to estimate the marginal posterior probability that contains a variable in the model. Hence the special case of the selection of the model is to choose a variable [62]. Linear Regression model, a most widely used statistical tool, is used for studying high-dimensional statistics. These includes ridge regression which permits to select simultaneously highly correlated covariates with same coefficients, Lasso regression which includes Group Lasso where preset group of covariates are selected together and Fused Lasso [63]. The Bayesian selection procedures which are used traditionally are more diplomatic than the prior options. There are few standard methods implemented such as a simple prior distribution, to select prior probabilities for the predictors, to prefer the posterior approach to ensure for predictors and estimation of posterior entities using MCMC sampling. Bayesian approach helps and furnishes model data combinations. The actual data values of the variables are maintained such that

3.6   Illustration with High-Dimensional Data

Head and Neck cancer, also called throat cancer, is defined as cancer that arise in the mouth, throat, nose, salivary glands, oral cancers, or other areas of the head and neck. Our study involved three datasets—hnscc, headnneck, and mcsurv.

1. hnscc: High-Dimensional gene expression data for head and neck cancer This data involved 565 patients, including 104 variables. It includes ID of patients, left censoring, death, overall survival, duration of progression-free survival, progression event, and high-dimensional covariates GJB1,...,HMGCS2.

2. headnneck: High-Dimensional Genomic data for head and neck cancer. It is a tibble of 13 columns including 467 subjects showing their overall survival, death, gender, stratum, and the arm of the group, along with six covariates $Y_1, ..., Y_6$.

3. mcsurv: Metronomic cancer data. It is a tibble of 15 columns, including 42 subjects showing their overall survival, death, time at which event occurred, and five variables, both on a discrete and continuous scale.

R packages contain a variety of resources, including R code, datasets, documentation, functions, and other supporting files. We had been successful in developing an R package named **SurviMCHd** which describes High-Dimensional Survival Data Analysis using Markov Chain Monte Carlo (MCMC). This R package is freely available through the Comprehensive R Archive Network

TABLE 3.1: Table showing Hazard rate with CI of variables

Function	Variables	HR	LCL	UCL
survintMC	GJB1	0.93	0.92	0.94
	PPP1RA	0.99	0.99	0.99
	HPN	1.00	1.00	1.00
	SLC4A4	0.84	0.83	0.85
	HNF1B	0.95	0.95	0.96

(CRAN) at `https://CRAN.R-project.org/package=SurviMChd` and runs on a variety of computing platforms. The user manual is also available from this site. To install **SurviMChd** package, use the following standard R syntax
$R > install.packages(\text{"SurviMChd"})$
The data for head and neck cancer was analyzed and was packed in R and certain functions namely **survintMC**, **survMC**, **survMCmulti**, **survexp-multi**, and **survweibMC** were formulated.The results obtained with those functions are discussed below.

The function **survintMC** describes High-dimensional survival analysis by MCMC method for interval-censored data. This function performs survival analysis with the MCMC method on 'hnscc' data by computing survival interval with left and right censoring given. In this dataset, there are multiple variables, so the survival columns should be arranged as the column containing left censoring information and the column containing right censoring information labeled as 'Leftcensor' and 'Rightcensor', respectively. For example, if we decide to select five columns starting from the 7th to 11th column having variables GJB1, PPP1RA, HPN, SLC4A, and HNF1B, then this function will give the hazard rate along with the confidence interval of those selected variables. The Function **survMC** and **survMCmulti** describes the survival analysis using MCMC method for univariate variables and perform survival analysis using MCMC method by selecting multiple variables respectively. The function "survMC" gives the means, standard deviation, quantiles, and mean deviance for univariate variable. In this function the survival columns of the data should be as death=1 if died or otherwise 0 7.3. The functions **survexpMC** and **survweibMC** describes the exponential and Weibull distribution by using MCMC method. Primarily, exponential distribution is a widely used continuous distribution in survival analysis. It is used to model the time elapsed between the events. There is high-dimensional data associated with survival analysis with survival duration as OS and event information (as death=1 if died or 0 if alive).

The Weibull distribution is a special case of three-parameter exponential distribution. This distribution is applicable to time to event data, failure time model, and model with reliability. This distribution with β close to or equal to 1 have a fairly constant failure rate, indicative of useful life or random

TABLE 3.2:   (survMC and survMCmulti)

Function	Variable	Means	SD	2.5%	97.5%	DIC
survMC	x1	0.67	1.16	0.50	0.89	219.42
	x2	0.94	1.07	0.81	1.07	227.08
	x3	0.55	1.94	0.15	2.03	227.18
	x4	1.01	1.01	0.99	1.02	226.79
	x5	4.65	1.50	2.11	10.06	212.61
survMCmulti	x1	0.67	1.17	0.50	0.91	212.18
	x3	0.90	2.06	0.23	3.96	
	x2	0.95	1.08	0.82	1.09	
	x4	1.00	1.01	0.99	1.02	

TABLE 3.3:   (survexpMC)

survexpMC	mu.vect	sd.vect	(2.5% , 97.5%)	Rhat(n.eff)
β_1	7.22	0.57	(6.46,8.01)	2.24(3)
β_2	-0.20	0.04	(-0.25,-0.15)	1.02(10)
β_3	0.03	0.06	(-0.03,0.11)	4.45(2)
β_4	0.16	0.62	(-0.77,0.10)	2.70(3)
β_4	-0.31	0.14	(-0.53,-0.10)	2.56(3)
β_6	-0.00	0.20	(-0.38,0.21)	1.86(4)
Deviance	9988.42	93.85	(9878.31,10152.11)	2.03(3)
survweibMC	**mu.vect**	**sd.vect**	**(2.5% , 97.5%)**	**Rhat(n.eff)**
β_1	7.39	0.28	(7.00,7.78)	0.97(10)
β_2	-0.05	0.02	(-0.07,-0.02)	0.93(10)
β_3	-0.01	0.02	(-0.04,0.02)	1.03(10)
β_4	-0.03	0.19	(-0.28,0.25)	1.04(10)
β_5	-0.02	0.17	(-0.24,0.20)	0.90(10)
β_6	0.02	0.14	(-0.26,0.15)	1.09(10)
Deviance	10115.68	17.35	(10086.43,10136.82)	1.65(4)

failure. It generalizes the exponential model to include non-constant failure rate functions. The 3-parameter Weibull pdf is given by:

$$f(x) = \frac{\beta}{\eta} \left(\frac{x - \gamma}{\eta} \right)^{\beta-1} e^{-\left(\frac{x-\gamma}{\eta} \right)^{\beta}} \tag{3.9}$$

where $\beta > 0$, $\gamma > 0$, and $\eta > 0$.

Here, 'β' is shape parameter known as Weibull slope, 'η' is the scale parameter, and 'γ' is the location parameter. This distribution can be reduced to two-parameter when location parameter '$\gamma=0$' and further to one-parameter when shape parameter '$\beta=c$' where c is a constant. Survival analysis studies the time to event data which is usually subject to censoring due to the termination of study. It indicates the relationships between the time to event data and also explains the several explanatory covariates [64]. Primarily it focuses on—whether an individual suffers a event or not and the time at which individual are examined at risk for outcome of interest. In many applications, models with more dimensions as compared to available sample size is frequent. Various different models are developed to deal with high-dimensional data in survival analysis. High-dimensional data deals where number of covariates are more than the independent samples are quite ubiquitous. In this research study, Bayesian survival analysis is studied particularly for high-dimensional data using **SurviMChd** software package. Bayesian analysis produces inference based on the information from the observed data and previous knowledge. This package involves the study of high-dimensional data for Bayesian Survival analysis using MCMC method. This study emphasizes high-dimensional data for Bayesian Cox Proportional model. In this paper, Bayesian Cox proportional hazard model for high-dimensional data is proposed for time to event data using various forms. These are related with high-dimensional data using bayesian approach which involved MCMC method, competing risk and also using exponential family of distribution like exponential and Weibull distribution. The Bayesian Cox proportional model is intermediate between the survival time and the covariates. Bayesian approaches have explicitly developed and became an impressive application with increasing power of model [65].

Bayesian modeling deals with distribution theory, hierachical models, networking diagrams, Markov Chain Monte Carlo (MCMC), and more. It counts for a risk factor of patients undergoing treatment which is termed as hazard ratio.The Bayesian Cox PH model is same as multiple linear regression method in which correspondence between a hazard and independent explanatory covariates for a particular duration of time. It also uses relative risk function. This approach can be computed feasibly even in high dimensions. In real-life applications, to deal with high-dimensional data this is quite useful which can give accurate results. In high-dimensional data, number of features exceeds the number of variables. The methods involved in this paper definitely be used for smooth implementation [65]. We also incorporated MCMC to deal with high-dimensional data for exponential and Weibull distribution, interval censored data with univariate and multivariate variables. This paper highlights study of high-dimensional survival data analysis using MCMC method.

3.7 Cox Proportional Hazard Ratio and Diagnosis

The Cox proportional hazards ratio (HR) is a measure of the relative risk of an event occurring in one group compared to another group. It is commonly used in survival analysis to compare the hazard of an event, such as death or disease progression, between different treatment groups or subgroups within a population. A HR greater than 1 indicates that the hazard of the event is higher in the first group compared to the second group, while a HR less than 1 indicates that the hazard of the event is lower in the first group compared to the second group.

In mediation analysis, a Cox proportional HR can be used to examine the relationship between a treatment or exposure and an outcome, with the mediator variable being the diagnosis status. This allows for the examination of whether the effect of the treatment or exposure on the outcome varies depending on the diagnosis status.

For example, a Cox proportional HR could be used to examine the relationship between a new drug and survival in a group of cancer patients, with the mediator variable being the diagnosis status (e.g. early-stage vs. advanced-stage cancer). This would allow for the examination of whether the effect of the drug on survival varies depending on the diagnosis status.

This type of analysis can provide valuable insights into the underlying mechanisms of the relationship between treatment or exposure, outcome, and the mediator variable, and inform the development of targeted interventions to improve outcomes for patients with different diagnosis status.

3.8   Stratified Cox Proportional Hazard Test

The Cox proportional hazards model, also known as the Cox model or the proportional hazards regression, is a type of survival analysis that is commonly used to analyze time-to-event data. It can be used to examine the relationship between a set of predictor variables and the time to a specific event, such as death or disease progression.

Stratified Cox proportional hazards test is an extension of the Cox proportional hazards model, where the sample is divided into subgroups or strata based on one or more variables. This approach allows for the examination of the relationship between the predictor variables and the time to event within each stratum separately.

In Mediation analysis, a stratified Cox proportional hazards test can be used to examine the relationship between a treatment or exposure and an outcome, with the mediator variable as a stratifying variable. This allows for the examination of whether the effect of the treatment or exposure on the outcome varies depending on the level of the mediator variable.

For example, a stratified Cox proportional hazards test could be used to examine the relationship between a new drug and survival in a group of cancer patients, with the mediator variable being the level of a specific biomarker. This would allow for the examination of whether the effect of the drug on survival varies depending on the level of the biomarker.

This type of analysis can provide valuable insights into the underlying mechanisms of the relationship between treatment or exposure, outcome, and the mediator variable, and inform the development of targeted interventions to improve outcomes.

3.9 Regression with Mediation Analysis

Regression analysis is a statistical method used to examine the relationship between a dependent variable and one or more independent variables. It can be used to predict the value of the dependent variable based on the values of the independent variables, and can also be used to determine the strength and direction of the relationship between the variables.

Mediation analysis is a statistical technique used to identify and quantify the mechanisms or pathways through which an exposure or treatment affects an outcome. It is often used in conjunction with regression analysis to determine the specific factors that mediate the relationship between the independent and dependent variables.

When combining regression analysis with mediation analysis, the goal is to identify the specific pathways through which an independent variable (i.e., the exposure or treatment) affects the dependent variable (i.e., the outcome). This can provide valuable insights into the underlying mechanisms of the relationship between the variables and inform the development of targeted interventions to improve outcomes.

In mediation analysis, the independent variable is regressed on the mediator variable and the dependent variable is regressed on both independent variable and the mediator variable. The difference in the coefficients of independent variable in these two regressions gives the indirect effect (i.e., mediated effect) of independent variable on the dependent variable through the mediator variable.

3.10 Survival Curve

Survival curves are a graphical representation of the probability of survival over time for a specific population or group. They are often used in medical research to compare the survival rates of different treatments or interventions, or to compare the survival rates of different subgroups within a population.

Mediation analysis can be used in conjunction with survival curve comparison to identify the specific pathways or mechanisms through which a treatment or intervention affects survival outcomes. The goal is to determine whether the treatment or intervention is having a direct effect on survival or whether the effect is being mediated through other factors, such as changes in health status or disease progression.

For example, in a clinical trial comparing a new drug to a placebo, a survival curve comparison would show the difference in survival rates between the two groups. Mediation analysis would then be used to determine whether the drug's effect on survival is due to a direct effect on the disease or whether it is due to changes in other factors such as blood pressure or inflammation.

This type of analysis can provide valuable insights into the underlying mechanisms of a treatment or intervention and inform the development of more effective strategies for improving survival outcomes.

3.11   Parametric Survival Analysis Using R

Parametric survival analysis is a method for modeling the time until an event of interest occurs, under the assumption that the data follows a particular probability distribution. The most commonly used distributions in survival analysis are the exponential, Weibull, and Gompertz distributions.

In R, there are several packages that can be used to perform parametric survival analysis, such as "survival", "flexsurv", and "parametric survival".

The "survival" package provides several functions for parametric survival analysis, including the "survreg()" function which can be used to fit parametric survival models with the exponential, Weibull, and Gompertz distributions. The function provides estimates of the parameters of the distribution, as well as the log-likelihood and residuals.

The "flexsurv" package provides a more flexible approach to parametric survival analysis. The "flexsurvreg()" function can be used to fit a wide range of parametric survival distributions, including the exponential, Weibull, and Gompertz distributions as well as more complex distributions such as the lognormal and loglogistic distributions.

It's important to note that parametric survival analysis requires specifying a distributional assumption, and it's recommended to consult with experts in survival analysis to ensure that the assumptions are appropriate for the data and the research question.

Exponential survival analysis with MCMC

```
data(headnneck)
survexpMC(m1=8,n1=12,m2=4,n2=7,chains=2,iter=10,
data=headnneck)
```

Result on exponential survival analysis with MCMC

	mu(SD)	2.5%	97.5%	Rhat	n.eff
beta1[1]	7.19(0.48)	6.36	7.78	2.28	3
beta1[2]	-0.10(0.06)	-0.20	-0.02	5.53	2
beta1[3]	-0.00(0.05)	-0.08	0.06	5.25	2
beta1[4]	0.19(0.29)	-0.20	0.68	2.30	3
beta1[5]	-0.23(0.25)	-0.63	0.17	2.65	3
beta1[6]	0.04(0.20)	-0.26	0.31	2.36	3
Dev	9820(45.3)	9867	9896	4.09	2

Weibull survival analysis with MCMC

```
data(headnneck)
survweibMC(m1=8,n1=12,m2=4,n2=7,
chains=2,iter=10,data=headnneck)
```

Result on weibull survival analysis with MCMC

	mu(SD)	2.5%	97.5%	Rhat	n.eff
beta1[1]	7.49(0.27)	7.07	7.82	1.48	5
beta1[2]	-0.05(0.02)	-0.08	-0.02	1.10	10
beta1[3]	-0.02(0.03)	-0.08	0.02	1.13	10
beta1[4]	-0.10(0.29)	-0.48	0.43	1.52	5
beta1[5]	0.07(0.13)	-0.10	0.26	0.99	10
beta1[6]	-0.09(0.22)	-0.36	0.26	1.07	10
Dev	10125(13.66)	10105.99	10145.60	1.58	5

Chapter 4

Competing Risk Modeling

4.1 Introduction

Classical survival analysis can be extended with Competing risks modeling. We can observe different events in addition to the occurrence of primary event of interest. Statistical methodology for analyzing competing risk modeling has rapidly changed over the last decades. However, clinical research commonly ignores the application of competing risk modeling. Sometimes, we forget them due to the need for more awareness and sometimes to the limitation of adopting complex applications. The aim of this chapter is to present the essential aspects of competing for risk modeling. We must describe the importance of the competing risk population of concern and highlight the increasing importance of competing risk modeling in clinical research. The entire survival analysis aspect becomes different by adopting competing risk modeling. For example, to present the incidence of death due to lung cancer among lung cancer patients, every patient will be followed from a baseline date (such as the date of diagnosis or date of surgery) until the date of death due to lung cancer or study closing date. A patient who dies of lung cancer during the study period would be considered to have an 'event' at their date of death. A patient who is alive at the end of the study would be considered to be 'censored'. Thus, every patient provides two pieces of information: follow-up time and status (i.e., event or censored). However, a patient can experience an event different from the event of interest. For example, a lung cancer patient may die due to causes unrelated to the disease. Such events are termed competing risk events.

4.2 Multistate Modeling of Competing Risks

A multistate model is used to model a process where subjects transition from one state to the next. For instance, a standard survival curve can be thought of as a simple multistate model with two states (alive and dead) and one transition between those two states. A diagram illustrating to make the

DOI: 10.1201/9781003298373-4

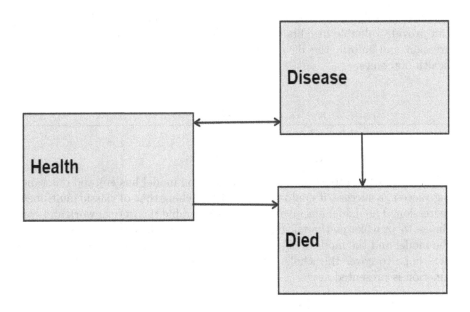

FIGURE 4.1: Multistate models illustration.

presentation. Particular study populations susceptible to competing risks are
(I) An individual is likely to host several yet unrecognized diseases, or they
are at risk of developing more than one disease that all may lead to clinically
relevant competing disease endpoints.
(II) The primary occurrence of such an endpoint strongly influences the individual. Examples are death, heart attack, cerebral stroke, etc.

4.3   Multistate Modeling in Mediation Analysis

Multistate modeling is a statistical technique used to analyze changes in
an individual's health status over time. It involves the use of mathematical
models to estimate the probabilities of transitioning between different health
states and can be used to predict future health outcomes.

Mediation analysis is a statistical technique used to identify and quantify
the mechanisms or pathways through which an exposure or treatment affects
an outcome. It is often used in conjunction with multistate modeling to determine the specific factors that influence the transition between different health
states.

When combining multistate modeling with mediation analysis, the goal
is to identify the specific pathways through which an exposure or treatment

affects the probability of transitioning between different health states. This can provide valuable insights into the underlying mechanisms of disease progression and inform the development of targeted interventions to improve health outcomes.

4.4   Proportional Hazards Models

Cox's (1972) proportional hazards regression model has enjoyed outstanding success, a success, it could be claimed, matching that of classic multilinear regression. The model has given rise to considerable theoretical work and continues to provoke methodological advances. Research and development into the model and the model's offspring have become so extensive that we cannot here hope to cover the whole field, even at the time of writing. The hazard function is presented as

$$\lambda(t|Z) = \lambda_0(t)\exp\{\beta Z\} \tag{4.1}$$

The parameter λ_0 is a fixed "baseline" hazard function. The term β is a relative risk parameter to be estimated. Now the term $Z = 0$ has the concrete interpretation with the baseline hazard $\lambda_0(t)$ as $\lambda(t|Z = 0) = \lambda_0(t)$. The vector of covariates is Z with the scalar product βZ that is interpreted as an inner product. It is common to replace the expression βZ by βZ where β and Z are $p \times 1$ vectors and b denoted the inner product of vectors with a and b.

4.5   Testing Covariates

In many medical research, the evaluation of treatment efficacy is based on the comparison of survival data between two treatment groups. This is often done through the ratio of two hazard functions of an endpoint event. Let T be the failure time of interest, V be the discrete cause of failure, and $Z(t)$ be the possibly time-dependent covariates. The conditional cause-specific hazards rate of cause k, $k = 1, ..., K$, given covariates $Z(t) = z$ is defined by

$$\lambda_k(t|z) = \lim_{\Delta t \to 0} \frac{1}{\Delta t} P(t \le T < t + \Delta t, V = k | T \ge t, Z(t) = z) \tag{4.2}$$

The function $\lambda_k(t|z)$ gives the instantaneous failure rate from cause k at time t in the presence of the other failure types. The conditional cumulative incidence function is given by $F_k(t|z) = P(T \le t, V = k|z)$. Both estimands are estimable from the competing risk data. Typical competing risks daa consist of independent and identically distributed replicates $X_i, \delta_i, \delta_i V_i, Z_i(.)), i =$

$1, ..., n$, of $(X, \delta, \delta V.Z(.)))$, where $X = \min(T, C), \delta = I(T \leq C)$. Here C is the censoring variable and δ is the censoring indicator. The cause of failure variaible V is observed if the failure time is uncensored, and is undefined otherwise. It is assumed that (T, V) and C are conditionally independent given covariates $Z(.)$. Many people have studied the problem of testing the equality of cause-specific hazards functions or cumulative incidence functions in the competing risks framework. The Cox model has been used to estimate the covariate effects on the cause-specific hazards function. If the covariates are time independent, the Cox model stipulates that the cause-specific hazard rates for different values of the covariates are proportional. A natural extension of the Cox model to address this situation allows for time-varying regression coefficients. The Cox model with time-varying regression coefficients has been studied earlier [66]. Recently, the model for the cause-specific hazard functions has been adopted to study the possibly time-varying for covariates [67].

4.6 Cause Specific Hazard Model

If there is one common category shared by all individuals at time origin, and if the other categories are absorbing, the longitudinal model is a time-discrete competing risks model and the conditional transition probabilities out of the initial category are time-discrete cause-specific hazards (CSHs). Recently, the approach to develop a parsimonious score testing procedure for investigating the impact of time-dependent high-dimensional covariates on 2 opposing clinical outcomes [68, 69]. Because of the high-dimensional multiple testing problem, the proportional odds assumption is attractive as it expresses the effect of one covariate in one regression coefficient. This approach is potentially also useful for general regression modeling of the CSHs of the competing risks. The reason is that standard regression approaches such as Cox models for all CSHs or Cox models for all so-called subdistribution hazards require as many models as there are competing event types to fully describe the competing risks process. Interpreting results of Cox models for all CSHs has been found to be potentially subtle, while interpretability of proportional subdistribution hazards, i.e. the Fine-Gray model for a cause-specific cumulative event probability, is subject of an ongoing debate [70].

4.7 Non-parametric Cause Specific Hazard Model

A Non-parametric Cause Specific Hazard Model is a type of survival analysis model that allows for the estimation of the hazard rate for different types of events (or causes) separately, without making any assumptions about the underlying distribution of the time-to-event variable. Cause-specific hazard models are used when there are multiple types of events that can occur, each with its own unique set of risk factors.

These models are non-parametric in the sense that they do not assume any particular functional form for the hazard function. Instead, they estimate the hazard function directly from the data. The most common approach to non-parametric cause-specific hazard modeling is to estimate the cause-specific hazard function using a non-parametric method such as the Kaplan-Meier estimator.

The Kaplan-Meier estimator is a widely used method for estimating the survival function in the presence of competing risks. It can be used to estimate the cause-specific hazard function by adjusting for the other causes of failure.

Non-parametric cause-specific hazard models are particularly useful when the goal of the analysis is to understand the risk of different events occurring, without making any assumptions about the underlying distribution of the time-to-event variable. They are widely used in fields such as medicine, epidemiology, and engineering to analyze time-to-event data.

4.8 Time-Dependent Covariates and Multistate Models

Time-dependent covariates and multistate models are statistical methods used to analyze time-to-event data when the risk of an event depends on the values of covariates that change over time. These methods are particularly useful when the goal of the analysis is to understand how the risk of an event changes as a function of time-varying covariates.

In time-dependent covariates models, the risk of an event is assumed to depend on the value of the covariates at the time of the event, rather than the value of the covariates at the time of the start of the study. This allows for the modeling of time-dependent changes in the risk of an event, such as changes in treatment or exposure status.

Multistate models are used when the outcome of interest is not a single event, but a sequence of events that occur over time. These models are used to estimate the transition probabilities between different states, such as the probability of transitioning from a healthy state to a diseased state or the probability of transitioning from a pre-disease state to a disease state.

Both time-dependent covariates and multistate models can be implemented using various statistical software such as SAS, R, or STATA and are widely used in fields such as medicine, epidemiology, and engineering to analyze time-to-event data. They provide a powerful tool for understanding how the risk of an event changes over time in relation to time-varying covariates and the sequence of events.

4.9   Competing Risk Analysis Using R

Competing risk analysis using R is a statistical method that uses the R programming language to perform competing risk analysis of time-to-event data. R is a widely used open-source programming language and environment for statistical computing and graphics. It has a rich ecosystem of packages and libraries that can be used to perform competing risk analysis.

There are several R packages that can be used to perform competing risk analysis, such as:

"cmprsk": This package provides a variety of functions for performing competing risk analysis, including the estimation of cumulative incidence functions and the calculation of subdistribution hazards. "survival": This package provides a wide range of survival analysis functions, including competing risk analysis. "competing.risk": This package provides a set of functions for the estimation of cumulative incidence functions and the calculation of subdistribution hazards. These packages allow for the estimation of cumulative incidence functions for different types of events and the calculation of subdistribution hazards. They also provide tools for model comparison, model checking, and visualization of the results. Competing risk analysis using R provides a flexible and powerful framework for analyzing time-to-event data with multiple possible events, and for making inferences about the underlying risk factors and their effect on the outcome.

In oncology and genomic research, the time to event is often observed as an outcome of interest. The time to event observations is measured as the date of registration to death or progression date. Sometimes, the duration can be measured from the date of treatment started instead of the registration date. Now it is different from calculating the duration as relapse or death of the advanced stage cancer patients. There could be two types of durations as measured, like date of registration to date of relapsed or up to date of death. Similarly, the event can be described as relapsed (yes or no) and died (yes or no). But statistical context becomes different while working with these two types of events jointly and separately. Now some patients may die without relapsed and some of them relapse. Before death, these different conditions provide the scope to work with the competing risk model.

The competing risk is defined based on p times latent variable. The survival function is presented as

$$S(t_1, t_2, ..., t_p) = P(T_1 > t_1, T_2 > t_2, ..., T_p > t_p) \qquad (4.3)$$

Now the subdensity for the event of type i is presented as

$$f_i(t) = (-\frac{\delta S(t_1, t_2, ..., t_p)}{\delta t_1})_{t_1 = t_2 = ... = t_p = t} \qquad (4.4)$$

It can be modified as survivor function for the event of type i as

$$S_i(t) = S(t_1 = 0, t_2 = 0, ..., t_i = t, ..., t_p = 0) \tag{4.5}$$

However, the subdistribution function presented as

$$F_j(t) = \int_0^t f_1(s)ds \tag{4.6}$$

It is defined as the probability of an event of type i by time t. It can present the sub-hazards with the bivariate algorithm as be shown that

$$h_i(t) = (-\frac{\delta\log(S(t_1, t_2, ..., t_p))}{\delta t_i})_{t_1 = t_2 = ... = t_p = t} \tag{4.7}$$

Similarly, the cause-specific hazard is presented by hazard function as

$$h_1(t) = -\frac{\delta\log(S_i(t))}{\delta t} = \frac{f_i(t)}{S_i(t)} \tag{4.8}$$

Now the p latent times are statistically independent as it is identical to the sub-hazard. Now the cause-specific hazard is similar to sub-hazard function. The cause-specific hazard is presented as the latent failure time approach of the competing risks.

4.10   Events in Data Analysis

A competing risk event refers to a situation where there are multiple types of events that can occur, each with its own unique set of risk factors. In such situations, the occurrence of one event may prevent the occurrence of another event. For example, in medical research, death due to a disease can be considered as an event, but death due to other causes can also be considered as an event. In this case, the death due to disease is the primary event of interest and the death due to other causes is considered as a competing event. In this case, the occurrence of death due to other causes would prevent the occurrence of death due to the disease. Competing risk analysis is used to account for this type of situation and to estimate the risk of different events occurring, taking into account the other possible events. Further, the competing risk was raised in the oncological domain while several types of consecutive events occurred like relapsed, progressed, metastasis, and deaths. Now the treatment regimes like chemotherapy, radiation, and surgery must have a different direct impact or combined on disease progression. However, for simplicity, we prefer to consider the single effect of the treatment on disease progression. But it is difficult to separate the impact of different types of treatment on the same patients and follow up disease progression. Further, the common type of disease progression is presented as death. The competing risk can be defined by sub-hazards, sub-density, subdistribution, and cause-specific hazard function.

4.11 Competing Risks in the Context of Bivariate Random Variable

Conventionally, the survival data are presented as bivariate random variables like (T, C). If the event has occurred, then it is defined as $C = 1$. If $C = 1$ then it is defined as the first member of the pair, T is the time at which the event occurred, and $C = 0, T$ while the observed measured as censored. Now it can be extended into the competing risks situation as $p \geq 2$ as types events as possible. Now the data can be presented as a pair (T, C) with the censoring indicator as C that is presented as 0, if the observation is censored. Sometimes, C can take value i as ordered statistics to present the initial failure/event measured as $(i = 1, 2, ...p)$. In cancer patients, if the relapsed occurred and after that, death appeared. The types of events can be presented in a different manner and presented by i.

4.12 Competing Risks with Latent Model

Similarly the competing risks can be presented through p latent, unobserved or as event times by $T_1, T_2, ..., T_p$. The time variable is presented as T and the mathematically it is presented as $T = \min\{T_1, T_2, ..., T_p\}$. If the censoring variable C is presented as $C = 0$ if the observation is censored and $C = i, i = 1, 2, ..., p$, otherwise. Sometimes, the T_1 can be presented by local relapse, T_2 time to metastatic, T_3 time to secondary malignancy, and T_4 time to death without any events; otherwise, it is zero.

4.13 Competing Risks as Bivariate Random Variable

Competing risks can be treated as a bivariate random variable, where the two variables represent the occurrence of two different types of events. A bivariate random variable is a set of two random variables that are defined over the same sample space. In the case of competing risks, the two random variables are the occurrence of the first event (failure) and the occurrence of the second event (competing failure).

The probability distribution of a bivariate random variable is usually represented by a joint probability distribution function or a joint cumulative distribution function. These functions describe the probability of both events occurring or not occurring together.

In competing risks analysis, the cumulative incidence function (CIF) is often used to estimate the probability of a specific event occurring, given that the other event has not occurred. The CIF can be used to estimate the probability of a specific event occurring, given that the other event has not occurred, and it can be represented as a function of the joint cumulative distribution function of the bivariate random variable.

Additionally, cause-specific hazard ratio (CSHR) is an important measure to study the effect of a covariate on the hazard of one event, given that the other event has not occurred yet. This measure can be calculated by using the bivariate survival function and the marginal survival function.

It's important to note that the bivariate random variable approach is useful when the goal is to understand the relationship between two types of events, but the data must be appropriately collected and the assumptions of the model must be verified. The CIF, or subdistribution, for an event of type $i(i = 1, 2, \ldots, p)$ is defined as the joint probability

$$F_i(t) = P(T \leq t, C = i) \tag{4.9}$$

In other words, the CIF is the probability that an event of type i occurs at or before time t. The overall distribution function is the probability that an event of any type occurs at or before time t. the overall distribution function is equal to the sum of CIFs, for all event types. Hence,

$$F(t) = P(T \leq t) = \sum_{i=1}^{p} P(T \leq t, C = i) = \sum_{i=1}^{p} F_i(t) \tag{4.10}$$

The subsurvivor function is the probability that an event of type i does not occur by time t and is defined as $S_i(t) = P(T > t, C = i)$. Note here that when the competing risks are not present, the overall distribution function spans the interval [0,1]. In contrast, in the competing risks environment the CIF can take values only up to $P(c = i)$ because

$$\lim_{t \to \infty} F_i(t) = P(C = i) \tag{4.11}$$

Therefore, $F_i(t)$ is not a proper distribution, hence the term Sub-distribution. Also note that

$$F_i(t) + S_i(t) = P(C = i) \tag{4.12}$$

In addition to the CIF and the subsurvivor function, the subdensity function for events of type i is defined as

$$f_i(t) = \frac{\delta F_i(t)}{\delta t} \tag{4.13}$$

As in the general setting of survival analysis (Chapter 2), the subhazard can be defined in mathematical terms as

$$\hat{h}(t)_i(t) = \lim_{\delta t \to 0} \{ \frac{P(t < T \le t + \delta t, C = i | T > t)}{\delta t} \} \tag{4.14}$$

The subhazard has the same interpretation as the hazard in the non-competing risks setting—the instantaneous event rate. The overall hazard of an event of any type can be found by summing over all subhazards:

$$h(t) = \sum_{i=1}^{p} \hat{h}_i(t) \tag{4.15}$$

It is worth pointing out that because the CIF is a joint probability, some of the relationships among the various subfunctions may not be as expected. Based on (2.3) one would expect that
$\hat{h}(t) = \frac{f_i(t)}{S_i(t)}$. However, simplifying (3.2) gives

$$\hat{h}(t) = \frac{f_i}{S(t)} \tag{4.16}$$

In contrast, the hazard function of the subdistribution (Gray, 1988) is defined as

$$\gamma_i(t) = \lim_{\delta t \to 0} \{ \frac{P(t < T \le t + \delta t, C = i | T > t \text{ or } (T \le \text{t and} C \ne i)}{\delta t} \} \tag{4.17}$$

The relationship between $\gamma_i(t)$ and the sub-density and the subdistribution can be expressed as

$$\gamma_i(t) = \frac{f_i(t)}{1 - F_i(t)} \tag{4.18}$$

The cumulative subhazard function is defined as

$$H_i(t) = \int_0^t h_i(x)dx = \int_0^t \{f_i(x)/S(x)\}dx \tag{4.19}$$

The material presented in this book is concerned primarily with nonparametric or semiparametric methods, which are more fully developed in the literature. However, parametric distributions are also used for modeling competing risks data, and several are considered here. Table 3.2 gives the functions defined in this section for several parametric distributions, when only two types of events are possible: type 1 is the event of interest and type 2 is the competing risk. This does not restrict the generality because all types of events other than the event of interest can be grouped together as a single competing risk.

4.14 Cumulative Incidence with Competing Risks

If the competing risk is not present then the relation between covariate on the hazard function and its effect on survival function can be established. If the event increases with time by hazard then effect of a covariate on the hazard function also increases. Commonly, the proportional hazard plays the role to work hazard function and it is required to fits the model with cause-specific hazard function along with the event of interest. The Fine and Gray provides the hazard model for the subdistribution hazard function for the any type of event as

$$\lambda_k^{sd}(t) = lim_{\Delta t \to 0} \frac{\text{Prob}(t \leq T \leq t + \Delta t, D = k | T \geq t \cup (T \leq t \cap D \neq k))}{\Delta t}$$

(4.20)

We can denote the type of events by D in the presence of K types of events. Sometimes, the hazard for a specific event is presented as the instantaneous appearance of the specific event among the individuals who are yet to experience the event. The Fine-Gray subdistribution provides the hazard model along with covariates on the subdistribution hazard function by

$$\lambda_k^{sd}(t|X) = \lambda_{k0}^{sd}(t)\exp(X\beta).$$

(4.21)

Here, $\lambda_{k0}^{sd}(t)$ gives the baseline subdistribution hazard function for the kth event type.

4.15 Statistical Software

There are different types of statistical softwares, and their abilities are different in performing the Fine-Gray subdistribution hazard regression model. Mostly, the used package in R is cmprsk, and the function crr is helpful to work with continuous time-varying covariates. Commonly, it is available to work with interactions of time-invariant covariates with known polynomial functions of time. It is free to consider the binary time-varying covariates. Sometimes, the risk regression functions are useful as a formula to work with the survival package. Thus, it is not feasible to consider the binary or categorical time-varying covariates toward working with competing risk modeling. Similarly, the stcrreg function available in Stata and PHREG available in SAS is suitable to work with categorical and continuous time-varying covariates. In addition, the external time-dependent covariates can also be incorporated in the Fine-Gray subdistribution hazard model. Initially, the external time-dependent covariates can be incorporated in the Fine-Gray subdistribution hazard model. However, it is required to carefully make inferences on regression estimates associated with cumulative incidence function (CIF). The statistical inference obtained on CIF needs careful consideration of the impact of the covariates. There are two ways to calculate the CIF, either by analytical or numerical integration, and it permits the comparison of the risks for subjects with different time-varying covariates. Now the simple ordering of the CIFs based on the sign of the regression coefficient in the proportional subdistribution hazard regression model. Similarly, extreme caution is required by incorporating the internal time-varying covariate in the Fine-Gray subdistribution hazard model. The subdistribution hazard makes it challenging to define the covariates for the individuals who have experienced a competing event. However, no one makes statistical inferences about the effect of the covariate on the CIF. There is a risk of getting misleading estimates of the subdistribution of the hazard ratio based on the assumption. The internal time-varying should be avoided while considering the Fine-Gray subdistribution hazard model. It suggests that their inclusion should be rare, and analysis should be performed very cautiously.

4.16 Discussion

The Fine-Gray subdistribution hazard model is popular for the time to event outcome research in the presence of the competing risk modeling. However, it requires careful consideration before allowing different time-varying covariates with the Fine-Gray subdistribution model. A similar challenge was also raised during work with Cox proportional hazard model. But accruing the competing risk model in the Cox proportional hazard model becomes tedious. Sometimes, we lose the ability to make statistical inferences to make the internal or subject-level time-varying covariates, and it is challenging [71, 72, 73]. Now the risk set contains the subjects who observed the competing event. Now the competing event incorporates death or cause-specific death. Then it will typically not be possible to know the value of the time-varying covariate after considering the individual has died, and ad hoc approaches for extrapolation are needed and complicating the interpretation of the model. Now competing risk model is extended toward multistate modeling toward developing the time-dependent covariate in a subdistribution hazards framework [72]. In this context, simulation is performed toward working with time-varying covariates in subdistribution hazard model [74]. However, the time-varying covariate had no association with the incidence of the primary outcome, as it is detected toward the non-null subdistribution hazard ratio. Now the avoiding the application of the subdistribution for approach for the effect of a time-dependent covariate [75]. But the application of internal and external time-dependent covariates and their conclusion should be determined by setting with the internal time-dependent covariates. For external covariates because CIF may be obtained from the subdistribution hazard model. Similarly, data without competing risks where the survival function can be obtained from the proportional hazard model. Fine-Gray model coupled with time-varying covariates is helpful to estimate the Cumulative Incidence [73]. Similarly, all types of cause-specific hazards can be accommodated by Cumulative Incidence [76]. It can be defined with $\{s_i : i = 1, \ldots, k\}$. This cumulative incidence can be prepared with competing risk survival along with time-varying covariate with k landmark times. Now ith landmark analysis can be performed for those who are at risk of the event at the ith landmark time. Now the inclusion of several landmark times declines the model consistency by increasing the variability. Suppose the prediction of incidence or survival probability is of our interest. In that case, this approach gives us more straightforward and more transparent with internal time-dependent covariates for the survival data without or with competing risks. The appealing feature of the Fine-Gray method is the accommodating capacity of the covariates on the CIF in the presence of competing risks, and it isn't easy to perform with conventional cause-specific hazard function.

Call survival library using R

Age	hgb	Stage	time	status
56	140	2	1.090	1
36	130	2	1.244	1
39	140	2	1.085	2
37	140	1	0.041	1
61	110	2	1.090	2
69	120	1	1.098	2
57	110	2	1.148	0
32	120	2	1.061	0
24	110	2	1.193	0
49	110	2	1.052	1

```
data(follic, package = "randomForestSRC")
follic.obj=rfsrc(Surv(time, status) ~ ., follic, nsplit = 3, ntree = 100)
plot.competing.risk(follic.obj)
```

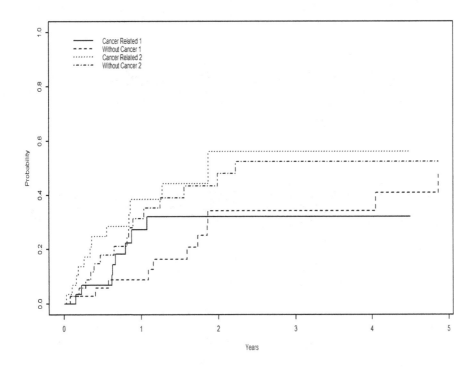

FIGURE 4.2: Competing risk plot using R.

unihdaftma funtion to obtain the competing risk analysis by univariate
accelerated failure time model with mediation analysis using R

```
data(hnscc2)
unihdaftma(m=8,n=80,survdur="os",event="death2",
ths=0.5,b=1000,d=10,data=hnscc2)
```

Died due to other causes/Progressed	No	Yes
No	258	0
Died	0	194
Died due to other causes	54	59

Died due to other causes/Died	No	Yes
No	258	0
Died	0	194
Died due to other causes	54	59

Died/Progressed	No	Yes
No	312	0
Yes	102	151

Selected variables obtained by unihdaftma function in autohd package using R

'Active variables and their beta and alpha means'
Active variable	beta.m	alpha.m
PCDP1	0.99	0.55
RSPH1	-0.71	0.98

---

hdraftma function to obtain the competing risk analysis using multivariate accelerated failure time model with mediation analysis

```
data(hnscc2)
hdraftma(m=8,n=100,survdur="os",event="death2",sig=0.1,
ths=0.02,b=10,d=10,data=hnscc2)
```

Selected variables obtained by hdraftma function in autohd package using R

'Active variables and their beta and alpha means'
Active variable	beta.m	alpha.m
AFF3	0.97	0.39
C3orf15	-0.74	0.98
CDHR5	-0.03	0.04

---

unicrma function to obtain the competing risk analysis by univariate accelerated failure time model with mediation analysis with Weibull distribution

```
data(hnscc2)
unicrma(m=8,n=100,survdur="os",event="death2",
sig=0.05,t=20,b=10,d=10,data=hnscc2)
```

Selected variables obtained by unicrma function in autohd package using R

[1] "No active variables"
"Number of active variables is 0"

Chapter 5

Accelerated Failure Time Modeling

5.1 Introduction

Globally, cancer's leading cause of death is contributed by almost 10 million deaths, and 19.3 million new cases only in 2020 [77]. Hence, rigorously improvement in treatment strategy and research is required to reduce cancer mortality. Commonly, mortality rate analysis is performed by the time to event survival analysis. The outcome variable is survival time, i.e., time until occurrence of the event or last follow-up. The treatment strategy can be improvised by the identification of the prognostic biomarkers in cancer. Broadly several worked carried out to identify biomarkers for cancer. However, progress in time-to-event data methodology is relatively more minor.

The survival analysis is broadly framed in three ways: if the pattern of survival time is taking as any particular distribution, it is parametric. If the exact distribution of survival duration is known, the parametric models are more appropriate than semiparametric and non-parametric models [78]. Parametric survival models recognize their consistency with theoretical survival function, complete specification with hazard and survival function, and time-quantile prediction [79]. If there is no information about the pattern of distribution, then the nonparametric model is suitable. Otherwise alternative is the Cox proportional hazard (CPH) model [49]. It is popular because of minimal assumptions.

A log-linear regression model is the Accelerated Failure Time (AFT) model, which is an alternative to the CPH model for statistical modeling in time-to-event data [80]. This model is mainly used for analyzing the censored clinical trial data. AFT model works on the assumption that the predictor variable's effect is accelerating or decelerating the life course of any disease (or interest of study) by some constant. At the same time, the CPH model works on the effect of the predictor variable by multiplying the hazard with some constant [81]. If the distribution pattern is known, then the parametric form of the AFT model with distributional choices provides a robust statistical inference than the hazard function [82]. It works on the premise of the effect of covariates, act proportionally (multiplicatively) concerning the survival time. The Bayesian approach is the alternative way to analyze the survival data, and it depends upon new knowledge or prior information of the experimental data [83]. The prior of the proposed method has been discussed in section 5.2. The challenges of the statistical methodology to work with high-dimensional survival data are not new. In contemporary statistics, to analyze the high-dimensional data, i.e., data with a small sample size (n) and a large number of covariates (p) is a significant issue [84]. Suppose a data has a large number of covariates. In that case, analyzing it with all covariates together is difficult and challenging, and variable selection techniques plays a vital role here. Some of existing variable reduction methods are LASSO [85], elastic net [86], Bayesian elastic net [87], Bayesian variable selection in AFT [88], weight function method for variable selection [89], feature selection method [90], etc.

The existing statistical literature on the AFT model and its generalized extension with other applications is vast. This manuscript presents some of those challenging works with the AFT model. Analyzing high-dimensional survival data is already a tough task in itself. In this article, we provide all the methodologies to estimate parameters of the AFT model for high-dimensional data with different features, such as conventional AFT model, AFT model using MCMC, and data with competing risk for both univariate and multivariate analysis. Various estimates for augmented data are also presented in this study. All these different types of methods support missing values in covariates. However, there are many R-packages to fit the AFT model, but they are not capable of working with the high-dimensional data with the AFT model using Bayesian. For example, R package bayesSurv [91] provides estimates of mixed-effects AFT model for censored data specification, package spBaysSurv [92] also includes AFT model for spatial and non-spatial survival data, package RobustAFT [93] is available for robust AFT regression for Gaussian and log-Weibull case, and function lss in package lss2 provides estimates for right-censored data in the AFT model which is based on the least-squares principle [94]. Therefore, it motivates us to create a methodological support that can work with AFT model for high-dimensional data along with variable selection. This article has explained a variable selection technique (using regularization) in the AFT model with MCMC. Variable or biomarker selection in any disease has been done before any predictive model (used AFT model with MCMC). The cue of this article is organized as section 5.2 describes the model and methods used. Section 5.3 is there for an explanation of the design and implementation of the package. Section 5.4 introduces the package afthd using example head and neck cancer data. Estimations using the proposed method(parametric AFT model using Bayesian on high-dimensional data with variable selection technique) have been done on two real examples (high-dimensional cancer data). For the validation purpose of methodology, simulation has been described in section 5.6. Developed R package afthd is available on the Comprehensive R Archive Network (CRAN) at https://cran.r-project.org/package=afthd.

5.2   Variable Selection

High-dimensional data, i.e., dataset generally have the dimension of predictor variables (covariates) is much larger than the sample size n. In that case, interpretation of data with all covariates is difficult, and hence reduction of unwanted covariates is needed. For parameter estimation purposes, ordinary least square or Maximum Likelihood Estimate(MLE) methods are used in a regression model that may be inclined more toward the training dataset and create overfitting problems. Hence, to overcome this over-fitting of the data, the regularization technique comes in to reduce the error and for tuning the model(function) by adding a penalty term in the error function. It has three types (i) L1 or LASSO, (ii) L2 or ridge, and (iii) elastic net regularization. In case of a penalty, the term $\lambda\Sigma_i|\beta_i|$ is added to the sum of squared residuals. Then it is said to be Least Absolute Shrinkage and Selection Operator(LASSO) or L1 regularization. Here λ is constant, and β is the parameter estimate. Let R_L is the lasso penalized function, and Y_e denotes the sum of squared residuals and p is the number of covariates, then it is expressed as

$$R_L = Y_e + \lambda\Sigma_i|\beta_i| \; ; \quad i = 1, 2, ..., p \tag{5.1}$$

If $\lambda = 0$, then equation (5.1) becomes the sum of squared residuals for the ordinary least squared method.

The penalty term $\lambda\Sigma_i\beta_i^2$ is added to the sum of squared residuals, and it is said to be Ridge or L2 regularization. Here λ and β have the same meaning as in LASSO regularization. Let R_R be the ridge penalized function, and Y_e denotes the sum of squared residuals, then it is presented as

$$R_R = Y_e + \lambda\Sigma_i\beta_i^2 \; ; \quad i = 1, 2, ..., p \tag{5.2}$$

Similarly, for $\lambda = 0$, then equation (5.2) becomes the sum of squared residuals for estimation using the ordinary least squared method.

If both penalty terms of L1 and L2 regularization are added to the sum of squared residuals, then it becomes elastic net regularization. If R_{en} be the elastic net penalized function and Y_e denotes the sum of squared residuals, it can be shown as

$$R_{en} = Y_e + \alpha\Sigma_i|\beta_i| + (1 - \alpha)\Sigma_i\beta_i^2 \; ; \quad i = 1, 2, ..., p \tag{5.3}$$

In equation, (5.3), if $\alpha = 0$ then it becomes ridge, if $\alpha = 1$ then it provides the result as LASSO and if $0 < \alpha < 1$ then this equation (5.3) is said to be elastic net regularization [86].

TABLE 5.1:   Common baseline distributions

Baseline Distribution	$f_\epsilon(x)$	$F_\epsilon(x)$	$W \equiv \exp \epsilon$
Normal	$\frac{1}{\sqrt{2\pi}}e^{-0.5x^2}$	$\Phi(x)$	Log-normal
Extreme Value	$e^x \exp(-e^x)$	$1 - e^{-\exp x}$	$Weib(1,1)$
Logistic	$\frac{\exp x}{(1+\exp x)^2}$	$\frac{\exp x}{1+\exp x}$	Log-Logistic

5.3   Accelerated Failure Time Models

This scenario is presented for high-dimensional time to event data having n number of individuals (sample size) and p covariates. All covariates are denoted as a vector $x : \{x_1, x_2, ..., x_p\}$. The Log-linear regression model for survival time $T_1, T_2, ..., T_n$ is taken as the AFT model and expressed as

$$log(T_i) = \beta_1 + \beta_2 x_{i1} + \beta_3 x_{i2} + ... + \beta_p x_{i(p-1)} + \beta_{p+1} x_{ip} = x_i'\beta + \sigma\epsilon_i \; ; \; i = 1, 2, ..., n \tag{5.4}$$

The term $log(T_i)$ denotes log-transformed survival time, β_1 is intercept, $(\beta_2, \beta_3, ..., \beta_{p+1})$ are regression coefficients corresponding to covariates $(x_1, x_2, ..., x_p)$ respectively, $\sigma = 1/\sqrt{\tau}$ is scale parameter, $\epsilon_i \overset{iid}{\sim} F_\epsilon(.)$. Here F_ϵ is known cumulative density function (CDF), defined on the real line for corresponding density f_ϵ and hazard h_ϵ [95]. It is shown in Table 5.1 for log-normal, Weibull, and log-logistic distributions in the AFT model. When equation (5.4) holds, it is expressed as

$$T_i \overset{ind}{\sim} AFT(F_\epsilon, \beta, \tau | x_i) \tag{5.5}$$

AFT model is based on the assumption that covariates act proportionally (multiplicatively) with respect to the survival time with the assumption

$$s(t|x) = s_0(\exp(\beta'x)t) \; ; \; t \geq 0 \tag{5.6}$$

where $s(t|x)$ is the survival function at the time t for covariate x, $s_0(\exp(\beta'x)t)$ be the baseline survival function at time t, and $\exp(\beta'x)$ is the acceleration factor. The covariate effect is said to be decelerated if $\exp(\beta'x) > 1$, and the covariate effect is said to be accelerated if $\exp(\beta'x) < 1$.

When baseline distribution in (5.4) be Normal, Logistic or Extreme value then corresponding to these, is represented by the AFT model for Log-normal, Log-logistic or Weibull distributions, respectively.

Using model (5.4), the survival time T is expressed as

$$S(t|x, \beta, \tau) = Pr(T > t | x, \beta, \tau)$$

$$= Pr[\frac{log(T) - x'\beta}{\sigma} > \frac{log(t) - x'\beta}{\sigma} | x, \beta, \tau]$$

$$= Pr[\epsilon > \frac{log(t) - x'\beta}{\sigma} | x, \beta, \tau] = S_\epsilon(t)$$

$$= 1 - F_\epsilon[\frac{log(t) - x'\beta}{\sigma}] \qquad (5.7)$$

The corresponding probability density function is derived from this as

$$f(t|x, \beta, \tau) = \frac{1}{t\sigma} f_\epsilon[\frac{log(t) - x'\beta}{\sigma}] \qquad (5.8)$$

All stated equations in this section work for multivariable in high-dimensional data, i.e., for vector x, a set of covariates $x_1, x_2, ..., x_p$. Similarly, all these equations also work for a single covariate(univariate) when p becomes 1.

5.4   Log Normal AFT Model

Now for fitting the high-dimensional data as log-normal AFT model for multivariate analysis, ϵ_i of equation (5.4) has a standard normal distribution, and hence T_i is distributed as log-normal. From equation (5.8), its density function comes out to be

$$f(t|x, \beta, \tau) = \frac{1}{\sqrt{2\pi}} \frac{1}{t\sigma} \exp[-\frac{1}{2\sigma^2}(log(t) - x'\beta)^2]; \ t > 0 \qquad (5.9)$$

Furthermore, it can be represented as $T \sim LN(x'\beta, \sigma^2)$, i.e., when $log(T) \sim N(\mu, \sigma^2)$ for $T > 0$, where $\mu = x'\beta$.
The survival function of the normal distribution is [82]

$$S(\epsilon) = 1 - \Phi(\epsilon) \qquad (5.10)$$

and the distribution function of normal distribution and its cumulative hazard functions are

$$\Phi(\epsilon) = \frac{log \ t - x'\beta}{\sigma} \qquad (5.11)$$

$$H_\epsilon(\epsilon) = -log\{1 - \Phi(\epsilon)\} \qquad (5.12)$$

In this way, log-normal AFT model is derived for multivariable. It has been used in the function lgnbymv() and lgnbyuni() for multivariate and univariate posterior estimates respectively.

5.5   Weibull AFT Model

If ϵ_i of equation (5.4) has extreme value distribution, then T_i follows Weibull distribution, and in this case, equation (5.4) will be denoted as the Weibull AFT model. i.e., $T \sim Weib(\sqrt{\tau}, e^{-x'\beta\sqrt{\tau}})$. Its pdf and cdf have been shown in Table 5.1

Exponential distribution is a particular case of Weibull. When $\tau = 1$ i.e., $\sigma^2 = 1$ then

$T \sim Weib(1, e^{-x'\beta})$ is equivalent to exponential distribution.

To interpret regression coefficients simpler in functions of the depicted package, we are using modified extreme value distribution with median 0. Hence we redefine the extreme value cumulative distribution function as

$$F_\epsilon(x) = 1 - \exp[-log(2)e^x] \qquad (5.13)$$

So we are using Weibull distribution that leads to AFT model when

$$T \sim Weib(\sqrt{\tau}, log(2)\exp(-x'\beta\sqrt{\tau})) \qquad (5.14)$$

We use these formulations to estimate the regression coefficient, scale parameter, shape parameter, and survival time. [96] has also used the Weibull AFT regression model to estimate survival time. A similar approach is used for univariate estimation.

5.6   Log Logistic AFT Model

Considering model (5.4) i.e., $T_i \sim AFT(F_\epsilon, \beta, \tau | x_i)$ with logistic error distribution. If baseline distribution is logistic, then T_1 follows log-logistic distribution, and equation (5.4) is said to be Log-logistic AFT model. Here it is expressed as

$$T \overset{ind}{\sim} LogLogis(x'\beta, \sqrt{\tau})$$

which is equivalent to

$$log(T) \overset{ind}{\sim} Logis(x'\beta, \sqrt{\tau})$$

and corresponding cdf (using the arguments in equation (5.7)) is :

$$F_\epsilon[(log(t) - x'\beta)\sqrt{\tau}] = 1 - \frac{1}{1 + \exp[(log(t) - x'\beta)\sqrt{\tau}]} = 1 - S(t|x, \beta, \tau) \quad (5.15)$$

In this way, using these functions, the log-logistic AFT model estimates parameters for multivariable and univariate covariates. We use these forms of AFT, i.e., log-normal, Weibull, and log-logistic, to analyze the high-dimensional data for multivariable and univariate cases using MCMC with variable selection technique. *AFT model with smooth time functions*

Functions have been designed for the AFT model without Bayesian, i.e., a conventional approach using the smooth time functions. Generalized survival models deliver a general and flexible approach to analyze the clinical-trial data. Survival function $S(t|x_i)$ for covariates x_i and to time t modeled as an AFT model by

$$S(t|x_i) = S_0(t \exp(-\eta(x_i, t; \beta))) \tag{5.16}$$

where β is the regression parameter and η is a function of t and covariates x. If AFT is a model with a time-constant accelerated factor, in that case, η is the linear predictor. The baseline survival function is modeled as

$$S_0(t) = \exp(-\exp(\eta_0(log(t); \beta_0))) \tag{5.17}$$

where η_0 is linear predictor, hence combined regression model is defined as

$$S(t|x) = \exp(-\exp(\eta_0(log(t) - \eta(x, t; \beta); \beta_0))) \tag{5.18}$$

Corresponding to it, the hazard function can be calculated as

$$h(t|x) = \frac{\partial}{\partial t}[-log(S(t|x))]$$

$$= \exp[\eta_0\{log(t) - \eta(x, t; \beta); \beta_0\}][\eta_0'\{log(t) - \eta(x, t; \beta); \beta_0\} + \eta_0\{\frac{1}{t} - \eta'(x, t, ; \beta)\}] \tag{5.19}$$

Here η_0 is the linear predictor and modeled using natural splines, and η can be freely modeled, provided this linear predictor is a smooth function of time. For design matrix $X(t, x)$, linear predictor in model using as $\eta(x, t; \beta) = X(x, t)\beta$. Construction of the linear predictor considers flexibility in it, such that time effects are twice differential and smooth, which allows possible interaction between covariates and time [97]. Estimation of the model is being done by Maximum Likelihood Estimates (MLE) as we are interested in a fully parametric model.

5.7 Graphical Methods and Model Assessment

Graphical methods and model assessment are important tools for evaluating the fit of a log-location-scale (LLS) model to a given dataset. These methods can provide insight into whether the assumptions of the LLS model are reasonable and whether the model is a good fit for the data.

Some common graphical methods used in model assessment include:

Residual plots: These plots show the difference between the observed data and the fitted values of the model. If the model is a good fit, the residuals should be randomly distributed with no patterns or trends.

Normal probability plots: These plots show a histogram of the residuals and a normal probability plot on the same graph. The normal probability plot is a graph of the residuals plotted against the expected normal distribution. If the model is a good fit, the points on the plot should fall close to a straight line.

Q-Q plots: These plots show a scatter plot of the quantiles of the residuals against the quantiles of the normal distribution. If the model is a good fit, the points on the plot should be close to a straight line.

Leverage plots: These plots show the relationship between the predictor variables and the residuals. If the model is a good fit, the residuals should be randomly distributed with no patterns or trends.

Partial regression plots: These plots show the relationship between the predictor variables and the response variable. If the model is a good fit, the residuals should be randomly distributed with no patterns or trends.

In addition to graphical methods, other model assessment techniques can be used, such as the Akaike information criterion (AIC) and the Bayesian information criterion (BIC), which are methods of comparing the goodness-of-fit of different models.

It is important to note that no single graphical method or model assessment technique can provide a definitive answer as to whether a model is a good fit for a particular dataset. A combination of techniques should be used to get a well-rounded assessment of the model's fit to the data.

5.8   Inference for Log-Location-Scale

Inference for log-location-scale (LLS) models involves estimating the parameters of the log-normal distribution that best fit the data. This typically involves maximum likelihood estimation (MLE) which is the process of finding the values of the parameters that maximize the likelihood function of the data.

For LLS models, the likelihood function is based on the probability density function (PDF) of the log-normal distribution. The MLE estimates for the location and scale parameters can be found by solving the equations of the likelihood function. These estimates can be found analytically or using numerical optimization algorithms such as the Newton-Raphson method.

Once the MLE estimates for the parameters have been found, standard error and confidence intervals for the estimates can be calculated using the inverse of the observed information matrix (Fisher information matrix) or using a bootstrap method.

Additionally, Goodness-of-fit tests can be used to evaluate the quality of the fit of the model, such as the Anderson-Darling test, Kolmogorov-Smirnov test, or the Chi-squared test. These tests can be used to check whether the log-normal distribution is a reasonable model for the data.

In general, the inference for LLS models is a standard practice in statistical analysis, however, it is important to take into account the assumptions of the model and its applicability to the specific data.

5.9 Extensions of Log-Location-Scale Models

Log-location-scale (LLS) models are a class of models that are used to model the distribution of continuous data. These models are based on the assumption that the data follows a log-normal distribution and are often used to model data that is positively skewed and has a long tail.

There are several extensions of LLS models that are commonly used in different applications:

Generalized Log-Location-Scale (GLLS) Models: These models allow for the data to follow other distributions, such as the Weibull distribution, instead of just the log-normal distribution.

Mixture Log-Location-Scale (MLLS) Models: These models allow for the data to be modeled as a mixture of different log-normal distributions.

Log-Location-Scale Mixture (LLSM) Models: These models are similar to MLLS models, but they allow for the mixture components to have different scale and location parameters.

Log-Location-Shape (LLS) Models: These models allow for the data to follow a log-normal distribution with a non-constant shape parameter.

Zero-Inflated Log-Location-Scale (ZILLS) Models: These models are used to model data that contains a large proportion of zero values.

These extensions of LLS models can be useful in different applications where the assumptions of the basic LLS model may not hold. They allow for more flexibility in modeling the data and can result in better model fits and improved predictions.

5.10   Hazard-Based Models with AFT

Hazard-based models with Accelerated Failure Time (AFT) modeling are a combination of two different types of models: hazard-based models and AFT models. A hazard-based model is a type of risk assessment model that assesses the potential harm or damage that could be caused by a particular hazard, while AFT modeling is a type of survival analysis that is used to model the time-to-failure of a system or component.

When these two types of models are combined, the hazard-based model is used to identify potential hazards and evaluate the potential consequences of an incident occurring, while the AFT model is used to estimate the time-to-failure of a system or component under different hazard scenarios. This allows for a more comprehensive assessment of the risk of an incident occurring and the potential consequences of such an event.

In many cases, these models are used in industries such as chemical manufacturing, oil and gas, and nuclear power to evaluate the risk of an accident or incident occurring and the potential consequences of such an event. The goal is to identify and mitigate potential hazards in order to reduce the risk of an incident occurring and minimize the potential impact on human health and the environment.

5.11   Bayesian Inference

For estimating the unknown parameter vector θ, we use Bayesian inference with MCMC posterior simulation. Our approach, which we are using in the AFT model with Bayesian, is based on conditional median priors [98]. Here prior specification on which it is made is a collection of median responses, each corresponding to a particular covariate combination [95]. We need a joint prior such as $p(\beta, \tau)$ that induces a distribution on regression coefficient. Here standard diffuse prior is $p(\beta, \tau) \propto 1/\tau$ [99]. A standard alternative to set β as normal given τ and the marginal of τ be gamma, which is usually a conjugate prior when the baseline distribution is log-normal or normal (as both have the same likelihood functions). In the AFT model, the median of baseline survival distribution is 0, the covariate is x and the median m is defined as $\exp(x'\beta)$. In multivariate Informative priors which we choose satisfies $p(\beta, \tau) = p(\beta)p(\tau)$, and that is why a discussion about the marginals for β and τ is required. analysis, covariate vectors are taken as $\tilde{x}_i = (x_{i1}, x_{i2}, x_{i3}, ..., x_{ip})'$, and the corresponding median time to event is defined as $\tilde{m}_i = \exp(\tilde{x}_i'\beta)$. From prior distribution which is specified on $\tilde{m} = (\tilde{m}_1, \tilde{m}_2, ..., \tilde{m}_p)$, distribution on β is induced. We can also write $\tilde{m}$ as $\tilde{m} \equiv \exp(\tilde{X}\beta)$ where $\tilde{X}$ is defined as

$\tilde{X} = (\tilde{x}_1, \tilde{x}_2, ..., \tilde{x}_p)'$. Since here we assume that $\tilde{X}$ is nonsingular, we produce $\beta = \tilde{X}^{-1} \log(\tilde{m})$. We assume that all $\tilde{x}_i$ are sufficiently "far apart", which is reasonable for the assumption that knowledge of $\tilde{m}_i$ can be observed independently. Hence, if we specify $p_{\tilde{m}}(.) = \prod_{i=1}^{p} p_{\tilde{m}_i}(.)$ as prior which we set on the vector of medians then

$$p(\beta) \propto \prod_{i=1}^{p} p_{\tilde{m}_i}(e^{\tilde{x}_i' \beta}) e^{\tilde{x}_i' \beta} \qquad (5.20)$$

and Jacobian of the transformation is

$$\mid \frac{d}{d\beta} exp(\tilde{X}\beta) \mid = \mid [\tilde{x}_i' exp(\tilde{x}_i' \beta) : i = 1, \ldots, p] \mid$$

$$= \mid Diag\{exp(\tilde{X}_i' \beta)\} \tilde{X} \mid$$

$$= \prod_{i=1}^{p} exp(\tilde{X}_i' \beta) \mid \tilde{X} \mid$$

$$\propto \prod_{i=1}^{p} exp(\tilde{X}_i' \beta)$$

Here $\tilde{m}$ is working as the scale of the data, and only we require to specify one prior on $\tilde{m}$. This same prior can be used to obtain a prior on β for baseline distribution with a median of 0. It is a good way to make a single elicitation irrespective of which or how many such AFT models will be considered. We assume a Gamma(0.001,0.001) prior for τ for each, which is an approximation of $p(\tau) = 1/\tau$, is taken as a choice to simple use [95]. It has a mean 1, mode 0, and a variance of 1000.

5.12 Accelerated Failure Time Using R

Accelerated Failure Time (AFT) analysis is a type of survival analysis that is used to model the time-to-failure of a system or component. In R, there are several packages that can be used to perform AFT analysis, such as the survival package, the flexsurv package, and the flexsurvreg package.

The survival package in R provides the survreg() function which can be used to fit AFT models. The function takes a survival object as input and the model formula should include the time-to-failure variable as the response variable and the predictor variables as the explanatory variables.

For example, if the data is in a dataframe called 'data' and the time-to-failure variable is called 'time' and the predictor variable is called 'x', the AFT model can be fit using the following code:

```
library(survival)
model=survreg(Surv(time, status) ~ x, data = data)
```

The flexsurv package in R provides the flexsurvreg() function which can also be used to fit AFT models. This function also takes a survival object as input and the formula should include the time-to-failure variable as the response variable and the predictor variables as the explanatory variables.

For example, if the data is in a dataframe called 'data' and the time-to-failure variable is called 'time' and the predictor variable is called 'x', the AFT model can be fit using the following code:

```
library(flexsurv)
model=flexsurvreg(Surv(time, status) ~ x, data = data)
```

The flexsurvreg package in R provides additional features for AFT models, such as handling of multiple failure times and left truncated data.

Once the model is fit, the summary() function can be used to obtain the estimates of the parameters, standard errors, and p-values, and the predict() function can be used to predict the time-to-failure for new data.

Additionally, the coef() function, exp() function, and confint() function can be used to obtain the coefficient estimates and confidence intervals of the model respectively.

It is important to note that the AFT models assumes that the data follows an exponential or Weibull distribution, so it is important to check the assumptions of the model and the goodness-of-fit of the model using graphical methods and statistical tests before interpreting the results.

5.13   Semiparametric Multiplicative Hazards Regression Models Using R

Semiparametric multiplicative hazards regression models are a type of survival analysis model that are used to model the relationship between a set of predictor variables and the hazard rate of an event. These models are called "semiparametric" because they do not make assumptions about the specific functional form of the hazard rate, but rather model it non-parametrically. The "multiplicative" part of the name refers to the fact that the hazard rate is modeled as a product of a parametric component and a nonparametric component.

The parametric component of the model is typically a linear function of the predictor variables, while the nonparametric component is modeled using a flexible function such as a spline or a kernel function. The nonparametric component allows for the hazard rate to change over time or for the hazard rate to have different shapes for different levels of the predictor variables.

In R, the package rms provides the regsem() function which can be used to fit a semiparametric multiplicative hazards regression model. The function takes a survival object as input and the model formula should include the time-to-failure variable as the response variable and the predictor variables as the explanatory variables. Additionally, the regsem() function allows for the nonparametric component of the model to be specified using a variety of options, such as splines, kernel functions, and other smooth functions.

For example, if the data is in a dataframe called 'data' and the time-to-failure variable is called 'time' and the predictor variable is called 'x', the semiparametric multiplicative hazards regression model can be fit using the following code:

```
library(rms)
model=regsem(Surv(time, status) ~ x, dist = "exp", data = data)
```

Once the model is fit, the summary() function can be used to obtain the estimates of the parameters, standard errors, and p-values, and the predict() function can be used to predict the hazard rate for new data.

It is important to note that semiparametric multiplicative hazards regression models are more flexible than traditional parametric survival models, and they allow for modeling of more complex hazard rate functions. However, they also require more computational resources and may be less interpretable than parametric models.

5.14   Model Selection Criteria

Model selection criteria can include things like accuracy, speed, scalability, interpretability, complexity, data availability, and resource constraints. The criteria used to select a model will depend on the specific application and the desired outcomes. Some criteria that might be used in a particular application include the ability to make accurate predictions, the ability to interpret the model, the ability to scale the model to different data sets, the ability to handle varying degrees of complexity, the availability of data for training and testing, and the availability of resources to run the model. There are many varieties for model selection in Bayesian inference. We are using deviance information criterion (DIC) as a model selection approach, proposed by [100]. The DIC delivers an assessment about a penalty of model complexity and fitting of the model. The deviance statistic for different cases of parametric AFT model has been provided using the Bayesian paradigm, indicating which model is more appropriate. For model selection purposes, we have developed a function aftbybmv() in presenting the R package. It estimates parameters and deviance information for each model, such as log-normal, Weibull, and log-logistic AFT model, and displays the one having minimum deviance. More minor, the deviance for any model indicates a better-fitting model. All the functions available in package 'afthd', running with Bayesian for univariate or multivariate analysis, provides deviance information.

5.14.1   Competing Risk Method

Since we work with cancer patients, any individual can die due to other causes except cancer. If the patient died due to different reasons, it is known as competing risk (CR). Now datasets having a death column need to revisit by putting death=0 (alive or censored), 1 (died), and 2 (death due to other causes). We provide the provision to work with the AFT model in CR data by a parametric system using Bayesian. The parametric approach is expanded in multivariate (*wbyscrkm*) and univariate (*wbyscrku*) setup in high-dimensional data in the Weibull AFT model.

The selection criteria for a competing risk model can vary depending on the application and data. In general, it is important to consider the following factors: (I) The model should provide an adequate fit to the data. This can be assessed with an appropriate goodness-of-fit measure such as Akaike Information Criterion (AIC) or Bayesian Information Criterion (BIC). (II) The model should include only the important explanatory variables and not include unnecessary or redundant parameters. (III) The model should be as simple as possible while still providing an adequate fit to the data. (IV) The model should be easy to interpret and provide meaningful results. (V) The model should provide an accurate prediction of outcomes. This can be evaluated using measures such as the Area Under the Receiver Operating Characteristic Curve (AUC) or the Concordance Index (C-index). (VI) The most appropriate method for handling missing data depends on the nature of the data and the purpose of the analysis. Common methods for handling missing data include. (VII) Replacing missing values with the mean value of the variable. (VIII) Replacing missing values with the median value of the variable. (IX) Replacing missing values with the most common value in the variable. (X) Replacing missing values with the mean of the K-nearest values. (XI) Replacing missing values with multiple imputed values.

Model selection criteria refers to the metrics used to evaluate the performance of different models and choose the best one for the given task. Common model selection criteria include: (I) The proportion of correct predictions made by the model. (II) The proportion of true positives among all predicted positives. (III) The proportion of true positives among all actual positives. (IV) The harmonic mean of precision and recall. (V) The area under the receiver operating characteristic curve. (VI) The negative log-likelihood of the predictions. In survival data, complications often occur due to the presence of missing values in covariates. Missing values in data can occur due to records lost by some happenstances or design of experiments; when any variable in data is unmeasured due to mistakenly or purposefully reasons, or when respondent subjects fail to attend the scheduled follow-up or did not respond to specific questions. Missing values in covariates in this work are overcome by the imputation technique. Here, the imputation method is monitored as replacing the missing value of a particular covariate by the mean of that covariate (using its non-missing responses). Before estimation procedures, this missingness in covariates is fulfilled by monitoring the data as explained. All the functions in presenting the R package support missing data handling for the covariates and provides posterior and MLE estimates for different AFT models.

5.15   Package Design and Implementation

The afthd package is an R package created to provide efficient tools for analyzing time to event data. It provides functions for computing various statistics, such as the autocorrelation function, variance, and the Hurst exponent, which can be used to identify trends in financial data. We presented the R package afthd to estimate the parameters in the context of survival modeling for the parametric AFT model. It is prepared with fourteen functions, and those are listed in Table 5.2. All the functions compatible to work with Bayesian are using the R2jags package. It makes fast execution of operations even at high iterations of MCMC. Functions *pvaft* and *rglaft* are working for estimates of parametric AFT model with smooth time function with conventional approach or without MCMC (uses packages rstpm2 and glmnet), other functions offer estimates for AFT model using MCMC. Table 5.2 summarizes the layout or short detail about the functions of the depicted package. The package contains Head and neck cancer data named 'hdata', on which we will provide an overview using all the functions inbuilt in afthd.

All the functions in package afthd have been created for different-different purposes in the context of the AFT model. All functions support data with missing values in covariates. The syntax or command for these functions is explained in this section. Some common arguments used in functions such as m and n are starting and end column numbers of covariates respectively study high-dimensional data. 'STime' is the name survival duration in data. The name event status in data is 'Event'(in which 1 represents the occurrence of the event and 0 for censored). 'data' is high-dimensional gene expression data that contains event status, survival time, and set of covariates. 'alpha' is chosen value between 0 and 1 to know the regularization method. alpha = 1 for Lasso, alpha = 0 for Ridge, and alpha between 0 and 1 for elastic net regularization technique. 'nc' and 'ni' are the number of Markov chains and

TABLE 5.2: Layout of package 'afthd'

Function	Used method or function	Variable taken as	Description
pvaft	'aft' from rstpm2	univariate	Estimates of covariates (together) with restriction on p value with parametric AFT model with smooth time functions without MCMC.
rglaft	'cv.glmnet' from glmnet, 'aft' from rstpm2	univariate	Estimates of selected variable using regularization(lasso, elastic net or ridge) in parametric AFT model using smooth time functions without MCMC.
wbysuni	AFT with Weibull dist	univariate	Posterior estimates of AFT model with Weibull distribution using MCMC (for all chosen covriates).
wbysmv	AFT with Weibull distribution	multivariate	Posterior multivariate estimates of chosen covariates (maximum 5 at a time) for AFT model with Weibull distribution using MCMC.
lgnbyuni	AFT with log-normal	univariate	Posterior estimates of chosen covariates(all together) for AFT model with log normal $dist^n$ using MCMC.
lgnbymv	AFT with log-normal	multivariate	Posterior estimate of chosen covariates(max 5 at once) for AFT model with log normal $dist^n$ using MCMC.
lgstbyuni	AFT with log-logistic	univariate	Posterior estimates (all chosen together covariates) of AFT model with log logistic $dist^n$ using MCMC.
lgstbymv	AFT with log-logistic	multivariate	Posterior estimates of chosen covariates(max 5 at once) for AFT model with log logistic $dist^n$ using MCMC.
aftbybmv	Weibull, log-normal, log-logistic	multivariate	Posterior parametric estimate of AFT model with min. deviance(DIC) among Weibull, log-normal and log-logistic $dist^n$.
rglwbysu	'cv.glmnet' from glmnet, Weibull dist	univariate	Bayesian univariate estimates (all selected together) of AFT model for selected covariates using regularization method.
rglwbysm	'cv.glmnet' from glmnet, Weibull	multivariate	Posterior estimates of AFT model for selected covariates using regularization method (Estimates for maximum first 5 selected covariates).
wbyscrku	Weibull dist, Competing risk	univariate	Posterior estimates (for all chosen covariates) with competing risk using AFT model of Weibull $dist^n$.
wbyscrkm	Weibull distribution, Competing risk	multivariate	Posterior estimates (max. 5 covariates at a time) with competing risk using AFT model of Weibull distribution.
wbyAgmv	Weibull dist, Augmented data	multivariate	Posterior multivariate estimate of AFT model with Weibull distribution using MCMC that supports augmented data.

the number of MCMC iteration, respectively.

5.15.1   Implementation in R

Short descriptions and illustrations of all the functions and their interpretation are provided in this section. We are using the head and neck cancer dataset from the package afthd. The dataset contains the survival time of 565 patients, their death status, and a set of covariates. This dataset is named hdata, which will be loaded using the package.

```
library(afthd)
data("hdata",package="afthd")
```

Next, function *pvaft* is used to obtain the univariate estimate for the parametric AFT model using the smooth time function. It provides a conventional approach for providing estimates for selected covariates from column number m to n of the dataset. Here argument *p* is used to make a restriction on the p-value for selecting covariates. i.e., only those covariates whose *p*-value is less than or equal to *p* will be picked. By default, it takes $p = 1$, which means all covariates from *m* to *n* will be selected. For example, we are choosing covariates from column number $m = 13$ to $n = 100$. Suppose we want only those covariates whose p-value is less than or equal to 0.01, hence taking $p = 0.01$.

```
pvaft(m=13,n=50,  STime="os",  Event="death",  p  =  0.01,
data=hdata)
```

Result on pvaft using R

Variable	Estimate	SE	z.value	p-value
ZMAT1	0.11	0.03	2.91	0.003

The Estimate std.Error z.value *p*_value for ZMAT1 is 0.11. It obtained a data frame that contains estimates of the regression coefficient for selected covariates. The next function *rglaft* is also developed for the AFT model using smooth time functions, but here, covariates selection is made with regularization technique. This function provides univariate estimates for the selected covariates using the conventional approach. For the implementation of this function *rglaft*, we are choosing covariates from column number $m = 8$ to $n = 45$ and variable selection technique as LASSO regularization, hence taking $alpha = 1$.

```
set.seed(1000)
head(rglaft(m=8, n=45, STime="os", Event="death", alpha=1,
data=hdata))
```

Variable	Estimate	SE	z.value	p-value
ZMAT1	0.11	0.03	2.91	0.00
GAL3ST1	-0.13	0.05	-2.40	0.01
LRP2	-0.09	0.04	-1.95	0.05
NME5	-0.08	0.04	-1.79	0.07
XIST	-0.02	0.01	-1.68	0.09
C2orf72	-0.06	0.03	-1.55	0.11

It obtains the result as a similar pattern of *pvaft*. Hence all the notations have the same meaning. Here, one thing can be noticed, in both functions, *pvaft* and *rglaft*, covariates are arranged in increasing order of their *p*-value in the outcome. It means upper covariates are more significant than lower ones.

5.16 AFT Model Using MCMC

Functions to estimate parametric AFT model using MCMC have been developed for log-normal, Weibull, and log-logistic distributions. All functions are available in the package for univariate estimates, display results for all chosen covariates at a glance. Whereas existing operations for multivariate estimation display estimates for the maximum first five selected covariates. All functions support missing values in covariates. Now, two functions are available in the package to analyze the data using the log-normal AFT model with MCMC. First, *lgnbyuni* for univariate estimation and *lgnbymv* for multivariate analysis. Execution of both procedures has been shown below for chosen covariates from column *m* to *n*, and their outcomes can be displayed by calling the assigned objects *lgn1* and *lgn2*.

Bayesian univariate analysis of AFT model with log normal using R

```
lgn1=lgnbyuni(m=10,n=11, STime="os", Event="death", nc=4,
ni=1000,data=hdata)
lgn2=lgnbymv(m=10,n=12, STime= "os", Event="death", nc=4,
ni=1000,data=hdata)
```

Result on Bayesian univariate analysis of AFT model with log normal by lgn1 using R					
Variable	Mean-coef(SD)	2.5%, 97.5%	Rhat	n.eff	Dev
HPN	-0.005(0.07)	[-0.15,0.14]	1.00	840	4050
SLC4A4	-0.065(0.07)	[-0.21,0.08]	1.00	2000	4049

Result on Bayesian univariate analysis of AFT model with log normal by lgn2 using R				
Variable	mean(sd)	2.5%, 97.5%	Rhat	n.eff
beta[1]	7.26(0.08)	[7.08,7.44]	1.00	280
beta[2]	0.01(0.08)	[-0.13 ,0.16]	1.00	490
beta[3]	-0.06(0.08)	[-0.22,0.10]	1.00	1500
beta[4]	-0.03(0.07)	[-0.18,0.12]	1.00	2000
Dev	4050(20.97)	[4008,4092]	1.00	310
sigma	1.57(0.07)	[1.44,1.73]	1.01	270
tau	0.40(0.03)	[0.33,0.48]	1.01	270

In the outcome, *lgnbyuni* provides estimates of the regression coefficient for listed covariates in the first column of the data frame. Whereas *lgnbymv* delivers estimates, in which $\beta 1$ is for intercept, and other β's for regression coefficients of covariates (in the order of selected covariates). Here, *sigma* is the scale parameter of the distribution, *Rhat* provides the convergence diagnostic, *n.eff* gives the efficient number of samples, and deviance obtains deviance Information criteria (DIC) of the model. Credible intervals and standard error are also available in the result.

For estimating the parameters using the Weibull AFT model with MCMC, function *wbysuni* is developed for univariate estimation and *wbysmv* is for multivariate. For example and to execute both these functions, we have chosen covariates from $m = 15$ to $n = 16$. It is shown below. Results have been assigned in objects *wb1* and *wb2*, which can be seen by calling it. The result of both functions contains a data frame of estimates of regression coefficients, their credible intervals.

```
wb1=wbysuni(m=15,  n=16,STime= "os", Event= "death", nc=4,
ni=1000,data=hdata)
wb2=wbysmv(m=15, n=16, STime= "os", Event= "death", nc=4,
ni=1000,data=hdata)
```

Variable	coef(SD)	[2.5%, 97.5%]
RGL3	-0.008(0.06)	[-0.13,0.11]
LMO3	-0.06(0.06)	[-0.18,0.07]

Variable	mean(sd)	[2.5%, 97.5%]	Rhat	n.eff
alpha	0.92(0.04)	[0.83,1.02]	1.01	360
beta[1]	7.28(0.07)	[7.14,7.42]	1.01	1300
beta[2]	0.02(0.07)	[-0.11,0.69]	1.01	200
beta[3]	-0.06(0.07)	[-0.20,0.85]	1.00	290
Dev	4052(21.27)	[4012,4.093]	1.01	1300
sigma	1.08(0.05)	[0.97,1.19]	1.01	360
tau	0.86(0.08)	[0.70,1.05]	1.01	360

Now for analyzing the high-dimensional data using log-logistic AFT model with MCMC, *lgstbyuni* is available in the package for univariate estimation and *lgstbymv* for multivariate. For example, choosing covariates from column number $m = 55$ to $n = 56$ to understand the working feature of these functions. Output in a data frame can be displayed by calling assigned objects *lgst1* and *lgst2* for *lgstbyuni* and *lgstbymv*, respectively. All the variables in output have the same meaning as explained before.

```
lgst1=lgstbyuni(m=55,n=56, STime="os", Event="death", nc=3,
ni=1000,data=hdata)
lgst2=lgstbymv(m=55, n=56, STime="os", Event="death", nc=3,
ni=1000,data=hdata)
```

lgst1

Variable	coef(Sd)	[2.5%, 97.5%]	Rhat	n.eff
KCNK3	-0.13(0.07)	[-0.28,0.01]	1.00	470
RGN	0.03(0.07)	[-0.11,0.18]	1.00	1500

lgst2				
Variable	mean(sd)	[2.5%, 97.5%]	Rhat	n.eff
beta[1]	7.22(0.08)	[7.07, 7.39]	1.00	610
beta[2]	-0.17(0.08)	[-0.34, -0.02]	1.00	670
beta[3]	0.09(0.08)	[-0.06, 0.25]	1.00	1500
Dev	962(21.73)	[922, 1006]	1.00	570
sigma	0.87(0.04)	[0.78, 0.95]	1.00	1300
tau	1.32(0.14)	[1.08,1.61]	1.00	1300

A function *aftbybmv* is developed for multivariate posterior estimates using an appropriate model among Weibull, log-normal, and log-logistic(which has minimum deviance) AFT model. For example, we choose covariates as column numbers $m = 10$ to $n = 11$ from the head and neck cancer dataset available in the package. Execution of this function can be done as done below. Its result is stored in the object *aft* and this will be displayed by calling it. The result shows the log-logistic AFT model fits well than other models for this taken example.

```
aft=aftbybmv(m=10, n=11, STime= "os", Event="death", nc=3,
ni=1000, data=hdata)
```

Variable	mean(sd)	[2.5%, 97.5%]	Rhat	n.eff
beta[1]	0.008(0.08)	[7.06,7.38]	1.00	620
beta[2]	0.0006(0.07)	[-0.14,0.15]	1.00	890
beta[3]	-0.05(0.08)	[-0.21,0.09]	1.00	1500
Dev	963(21.00)	[923.84,1006]	1.00	1100
sigma	0.87(0.04)	[0.79,0.96]	1.01	140
tau	13(0.13)	[1.06,1.59]	1.01	140

5.17   Weibull AFT Model with Different Specifications

Functions *rglwbysu* and *rglwbysm* are available in the package for univariate and multivariate posterior estimates, respectively. Both provide estimates using the Weibull AFT model for those covariates selected by regularization technique from chosen covariates m to n. Here, for example, initially, we are selecting covariates as column numbers $m=10$ to $n = 22$ and choosing $alpha = 1$, i.e., LASSO regularization technique for covariates selection. Objects *rglw1* and *rglw2* are assigned as a result of both functions. This implementation is shown below.

```
set.seed(1000) rglw1=rglwbysu(m=10, n=22, STime= "os", Event=
"death", nc=3, ni=1000, alpha=1, data=hdata)
rglw2=rglwbysm(m=10, n=22, STime= "os", Event= "death", nc=3,
ni=1000, alpha=1, data=hdata)
```

For analyzing the high-dimensional data with competing risk, two functions, *wbyscrku* (for univariate) and *wbyscrkm* (for multivariate) are developed and available in the package. Both functions provide posterior estimates using the Weibull AFT model. For example, we choose covariates from m to n in dataset *hdata* as column numbers to implement both functions. Results of both functions are stored in objects *wcrk1* and *wcrk2* will be displayed by calling it. As event status in these functions contains 0 for censored, 1 for death due to interest of study, and 2 for the event's occurrence due to other causes, the outcome delivers estimates for two different cases separately. One for event status (0,1) and another for (0,2). Here 0 (censored) is taken as reference.

```
wbyscrku(m=10, n=11, STime= "os", Event= "death2", nc=3,
ni=1000, data=hdata)
wbyscrkm(m=10, n=12, STime= "os", Event= "death2", nc=3,
ni=1000, data=hdata)
```

There is a function named *wbyAgmv* in the package, which provides multivariate posterior estimates for covariates as well as an estimate of survival time after t days/months/years (same unit as survival time) for augmented data (i.e., augmented rows are already added in last rows of the dataset). Here one more advantage is, the estimate of survival time can also be estimated only for the desired number of individuals present in the row numbers p to q of the dataset. For example, we choose covariates from $m = 10$ to $n = 12$ and estimate survival time for individuals from p = 560 to q = 565. It is executed as shown below, and its result is stored in the object *aug* . It can be displayed by calling it.

```
wbyAgmv(m=10, n=12, p=563, q=565, t=200, STime="os",
Event="death",nc=3, ni=1000, hdata)
```

Variable	mean(sd)	[2.5%, 97.5%]
STime	0.89(0.01)	[0.86,0.9]
STime	0.88(0.02)	[0.84,0.92]
STime	0.89(0.01)	[0.86,0.91]
beta[1]	7.28(0.07)	[7.14,7.41]
beta[2]	-0.01(0.07)	[-0.15,0.13]
beta[3]	-0.02(0.07)	[-0.16,0.121]
beta[4]	-0.06(0.07)	[-0.19,0.08]
deviance	4050(20.85)	[4012,4089]
sigma	1.08(0.05)	[0.98,1.19]
tau	0.85(0.08)	[0.70,1.03]
Overall_S	0.89(0.01)	[0.86,0.92]

In the output section, 'STime' provides the estimated value of survival time 'os' of data for chosen individuals from row number p to q. *Overall_S* in output provides an overall estimate of survival time 'os' in data for all individuals (i.e., for nrow(data)). Other variables in the result have the same meaning as explained before.

5.18   Simulation Study

In this section, we are verifying the performance of methodology, i.e., log-normal, Weibull, and log-logistic distribution in AFT model using simulation. Here simulation study is done as estimating the posterior estimates with varying priors. We have taken the priors for β and τ as explained in subsection 2.3 and fixed as $\beta \sim norm(0, 0.000001)$ and $\tau \sim Gamma(0.001, 0.001)$ in models. Now to verify the accuracy of models, we obtained the posterior estimates of the first real example data with varying priors as shown in Table 5.2 (with varying β and fixed τ) and Table 5.3 (with varying τ and fixed β). For these Bayesian estimates, we run the code with 3 number of MCMC chains and 10,000 iterations. From these obtained posterior estimates, we can easily see the consistency of estimates of regression coefficients ($\beta2$ to $\beta6$) in all three distributions, log-normal, Weibull, and log-logistic of AFT model, and *Figure* 5.1 provides its easy visualization.. In the figure, visualization of posterior estimates is provided with one of τ and β are at fixed and another is varying. The first column of the figure is for Weibull distribution, the second is for log-normal, and the third column is for log-logistic. The first and second rows of the figure are for fixed τ and varying β at SD and mean, respectively. Similarly, the third and fourth rows are for fixed β and varying τ at SD and

TABLE 5.3: Estimation of coefficient of regression in simulated study with fix $\tau \sim Gamma(0.001, 0.001)$ and varying β, $\beta \sim Normal(\mu, \sigma)$

	Weibull	Log-normal	Log-logistic	Weibull	Log-normal	Log-logistic
	Mean (SD)	Mean (SD)	Mean (SD)	Mean (SD)	Mean (SD)	Mean (SD)
	$(\mu = 0, \sigma = 10^{-6})$			$(\mu = 1, \sigma = 10^{-6})$		
$\beta 1$	7.4(0.09)	7.44(0.11)	7.38(0.1)	7.38(0.09)	7.46(0.12)	7.39(0.11)
$\beta 2$	−0.26(0.1)	−0.27(0.11)	−0.26(0.11)	−0.24(0.09)	−0.27(0.11)	−0.27(0.11)
$\beta 3$	−0.19(0.08)	−0.23(0.1)	−0.21(0.09)	−0.18(0.07)	−0.23(0.1)	−0.22(0.09)
$\beta 4$	−0.22(0.06)	−0.29(0.09)	−0.26(0.08)	−0.22(0.06)	−0.3(0.09)	−0.26(0.08)
$\beta 5$	−0.25(0.1)	−0.23(0.11)	−0.25(0.1)	−0.25(0.09)	−0.24(0.11)	−0.25(0.10)
$\beta 6$	−0.21(0.08)	−0.26(0.09)	−0.26(0.09)	−0.20(0.07)	−0.27(0.1)	−0.26(0.09)
deviance	2595(15.5)	2621(14.5)	659(16)	2593(16.3)	2624(15.9)	661(16.4)
	$(\mu = 0, \sigma = 10^{-5})$			$(\mu = 10, \sigma = 10^{-6})$		
$\beta 1$	8.00(0.09)	7.42(0.12)	7.38(0.1)	7.38(0.09)	7.45(0.12)	7.38(0.11)
$\beta 2$	−0.26(0.09)	−0.26(0.11)	−0.27(0.12)	−0.25(0.09)	−0.26(0.11)	−0.27(0.11)
$\beta 3$	−0.18(0.08)	−0.23(0.1)	−0.22(0.09)	−0.18(0.07)	−0.23(0.1)	−0.22(0.09)
$\beta 4$	−0.22(0.07)	−0.3(0.09)	−0.26(0.08)	−0.21(0.065)	−0.3(0.09)	−0.26(0.08)
$\beta 5$	−0.23(0.09)	−0.23(0.1)	−0.26(0.1)	−0.25(0.09)	−0.24(0.12)	−0.25(0.1)
$\beta 6$	−0.20(0.07)	−0.27(0.09)	−0.26(0.09)	−0.21(0.07)	−0.27(0.09)	−0.27(0.09)
deviance	2593(16.4)	2623(15.5)	660(15.8)	2468(4)	2592(15.3)	660(16.5)
	$(\mu = 0, \sigma = 10^{-4})$			$(\mu = 100, \sigma = 10^{-6})$		
$\beta 1$	7.39(0.09)	7.45(0.12)	7.38(0.1)	7.38(0.08)	7.46(0.12)	7.39(0.11)
$\beta 2$	−0.26(0.1)	−0.27(0.11)	−0.274(0.11)	−0.26(0.09)	−0.27(0.11)	−0.27(0.11)
$\beta 3$	−0.18(0.07)	−0.23(0.1)	−0.22(0.09)	−0.18(0.08)	−0.22(0.1)	−0.22(0.09)
$\beta 4$	−0.21(0.06)	−0.3(0.09)	−0.26(0.08)	−0.22(0.07)	−0.3(0.09)	−0.26(0.09)
$\beta 5$	−0.25(0.09)	−0.23(0.11)	−0.25(0.1)	−0.24(0.1)	−0.24(0.11)	−0.25(0.1)
$\beta 6$	−0.21(0.07)	−0.26(0.09)	−0.27(0.09)	−0.21(0.07)	−0.26(0.1)	−0.26(0.09)
deviance	2593(15.3)	2623(16.2)	661(16)	2468(3.8)	2623(15.9)	661(16.1)
	$(\mu = 0, \sigma = 10^{-3})$			$(\mu = -1, \sigma = 10^{-6})$		
$\beta 1$	7.39(0.1)	7.43(0.11)	7.38(0.11)	7.39(0.09)	7.46(0.12)	7.39(0.11)
$\beta 2$	−0.26(0.09)	−0.26(0.11)	−0.27(0.11)	−0.27(0.1)	−0.28(0.11)	−0.27(0.11)
$\beta 3$	−0.18(0.07)	−0.22(0.01)	−0.22(0.09)	−0.18(0.08)	−0.23(0.1)	−0.22(0.09)
$\beta 4$	−0.22(0.06)	−0.29(0.07)	−0.26(0.09)	−0.22(0.06)	−0.30(0.1)	−0.26(0.08)
$\beta 5$	−0.24(0.1)	−0.24(0.11)	−0.25(0.1)	−0.24(0.09)	−0.23(0.11)	−0.25(0.1)
$\beta 6$	−0.21(0.07)	−0.26(0.09)	−0.27(0.09)	−0.20(0.07)	−0.26(0.09)	−0.26(0.09)
deviance	2594(17)	2620(16)	661(16.1)	2594(16.8)	2624(15.8)	661(16.5)
	$(\mu = 0, \sigma = 10^{-2})$			$(\mu = -10, \sigma = 10^{-6})$		
$\beta 1$	7.37(0.09)	7.45(0.11)	7.38(0.11)	7.39(0.09)	7.46(0.12)	7.39(0.11)
$\beta 2$	−0.25(0.09)	−0.26(0.111)	−0.27(0.11)	−0.25(0.1)	−0.27(0.11)	−0.27(0.11)
$\beta 3$	−0.18(0.07)	0.23(0.097)	−0.22(0.09)	−0.18(0.08)	−0.23(0.1)	−0.22(0.09)
$\beta 4$	−0.22(0.06)	−0.3(0.074)	−0.26(0.08)	−0.22(0.06)	−0.3(0.09)	−0.26(0.09)
$\beta 5$	−0.24(0.1)	−0.22(0.109)	−0.25(0.1)	−0.25(0.1)	−0.23(0.11)	−0.26(0.1)
$\beta 6$	−0.21(0.07)	−0.26(0.088)	−0.26(0.09)	−0.21(0.07)	−0.27(0.1)	−0.26(0.09)
deviance	2592(15.9)	2627(15.8)	660(16.3)	2594(15.6)	2623(15)	661(16.4)

TABLE 5.3: (*Continued*)

	Weibull	Log-normal	Log-logistic	Weibull	Log-normal	Log-logistic
	Mean (SD)	Mean (SD)	Mean (SD)	Mean (SD)	Mean (SD)	Mean (SD)
	($\mu = 0, \sigma = 10^{-1}$)			($\mu = -100, \sigma = 10^{-6}$)		
$\beta 1$	7.38(0.09)	7.43(0.111)	7.37(0.1)	7.38(0.1)	7.45(0.12)	7.38(0.11)
$\beta 2$	−0.27(0.09)	−0.27(0.11)	−0.27(0.11)	−0.26(0.09)	−0.27(0.11)	−0.27(0.11)
$\beta 3$	−0.18(0.07)	−0.23(0.097)	−0.22(0.09)	−0.18(0.08)	−0.23(0.1)	−0.22(0.09)
$\beta 4$	−0.21(0.06)	−0.3(0.074)	−0.26(0.08)	−0.22(0.06)	−0.3(0.09)	−0.26(0.08)
$\beta 5$	−0.24(0.1)	−0.23(0.108)	−0.25(0.1)	−0.25(0.09)	−0.23(0.11)	−0.25(0.10)
$\beta 6$	−0.20(0.07)	−0.26(0.087)	−0.26(0.09)	−0.20(0.07)	−0.26(0.1)	−0.26(0.09)
deviance	2592(15.8)	2620(15.4)	659(15.8)	2593(16.1)	2623(16.2)	660(15.9)

mean, respectively. Hence, it can be concluded that these posterior estimates with varying priors on the same distribution give a consistent result, and hence AFT model with these three distributions provide adequate performance.

5.19 Analysis on Two Real Gene Expression Data

In this section, we are analyzing two real liver cancer sides of gene expression datasets, which have been obtained from the National Cancer Institute's The Cancer Genome Atlas (TCGA) database. It provides both clinical and gene expression data from primary tumor patients samples for multiple cancer types. The first data has 1881 biomarkers, and the second data contains 18526 biomarkers. Using the following steps, we are analyzing these two data using the developed R package 'afthd'.

Step 1: First, we make variable (covariates here as biomarkers) selection using regularization technique, using function rglaft().

Step 2: Estimates for these selected covariates will be estimated using (i) conventional approach of AFT model, (ii) Bayesian AFT model for univariate, and (iii) Bayesian AFT model for multivariable.

Here three points are to be noted. First, to estimate the parameters of the parametric AFT model with the conventional approach for selected covariates using regularization. A single function *rglaft* will do all these procedures and provides the desired outcome. Second, for univariate estimation of the parameters of parametric AFT model with Bayesian for selected covariates using regularization is done by single-function *rglwbysu*. The third point for multivariate estimation of the parameters of the AFT model with Bayesian for selected covariates (first five covariates) using regularization is made by the single function *rglwbysm*. Here in both examples, there are 1881 and 18526 covariates. In the first step, we make a covariate selection as biomarkers. In variable selection, it is possible to get more than five covariates. Since it is well-known that multivariate estimates for more than five covariates do not explain well about the parameters. Hence to estimate the most significant five covariates, we take the first five uppermost covariates from the output of function *rglaft* (as it is in increasing order of *p*-value). Thereafter we use these covariates (with survival time and event status) in function *rglwbysm*, which provides its multivariate analysis using MCMC. Fourth, there is an advantage of all the functions of package afthd; it provides estimates for all covariates together and does not need to run the code for all the variables one by one.

The first TCGA data of liver cancer is specified as 1881 biomarkers (as covariates), event as death, survival time as Overall Survival (OS), and with 419 samples, out of these samples, 164 samples reached death during follow-up, and others are censored. Now we are analyzing this data (named as *dt*1) with the AFT model. We are first estimating with the conventional approach of parametric AFT model with variable selection regularization technique LASSO. It is shown obtained in Table 5.4.

For this same first data, Bayesian univariate estimates using the parametric AFT model for selected covariates with function *rglaft* are presented in Table 5.5.

Now multivariate posterior estimates for selected first five uppermost covariates from the output of the function *rglaft* (contained in data named 'd22') is done by *wbysmv*. In the covariate selection scenario, basically, we are applying a two-step variable selection criteria, first by LASSO and second by MLE (since covariates in the output of *rglaft* are arranged in increasing order of p-value). This whole process could be done in single-function *rglwbysm*, but here variable selection would be made only with LASSO. Hence for more significant estimation, we are doing in two steps. Multivariate posterior estimates for selected covariates are shown in Table 5.6.

There are 18526 biomarkers (as covariates) in the second TCGA data of liver cancer of sample size 216. In this time-to-event data, the event is specified as death and survival time as OS. Out of 216 samples, 66 samples reached the event, and other samples are censored. Analysis of this data (named as 'dt2') is done with the AFT model. First analysis and estimation with the conventional approach of parametric AFT model with variable selection LASSO for data with 18526 covariates are obtained in Table 5.7.

Different estimates of the regression coefficient for most significant variables (ILMN_1794643, ILMN_1699496, ILMN_1795183, ILMN_1693598, ILMN_2070052) selected by LASSO are obtained using function *wbysmv*. Let 'dt22' is data that contains these variables and demographic variables. Hence now its multivariate Bayesian estimate of the AFT model is shown in Table 5.8.

TABLE 5.4: Estimation of regression coefficient in simulated studies for fixed $\beta \sim Normal(0, 0.000001)$ and varying τ, if $\tau \sim Gamma(a, b)$

	Weibull Mean (SD)	Log-normal Mean (SD)	Log-logistic Mean (SD)	Weibull Mean (SD)	Log-normal Mean (SD)	Log-logistic Mean (SD)
	$(a = 0.001, b = 0.002)$			$(a = 0.002, b = 0.001)$		
$\beta 1$	7.38(0.09)	7.44(0.12)	7.38(0.1)	7.38(0.09)	7.45(0.13)	7.39(0.11)
$\beta 2$	−0.26(0.1)	−0.27(0.11)	−0.27(0.12)	−0.24(0.1)	−0.27(0.1)	−0.27(0.1)
$\beta 3$	−0.18(0.07)	−0.22(0.1)	−0.22(0.09)	−0.19(0.08)	−0.22(0.1)	−0.22(0.09)
$\beta 4$	−0.22(0.06)	−0.3(0.08)	−0.26(0.08)	−0.21(0.06)	−0.3(0.09)	−0.26(0.09)
$\beta 5$	−0.24(0.09)	−0.23(0.1)	−0.25(0.1)	−0.24(0.09)	−0.24(0.1)	−0.25(0.1)
$\beta 6$	−0.21(0.07)	−0.27(0.09)	−0.27(0.09)	−0.21(0.07)	−0.26(0.09)	−0.26(0.09)
deviance	2593(16.9)	2622(16)	661(15)	2594(16)	2622(16.4)	661(16)
	$(a = 0.001, b = 0.003)$			$(a = 0.003, b = 0.001)$		
$\beta 1$	7.38(0.09)	7.45(0.12)	7.39(0.11)	7.38(0.09)	7.45(0.11)	7.38(0.1)
$\beta 2$	−0.27(0.09)	−0.26(0.1)	−0.26(0.1)	−0.26(0.09)	−0.27(0.11)	−0.27(0.11)
$\beta 3$	−0.17(0.07)	−0.22(0.1)	−0.23(0.09)	−0.17(0.08)	−0.23(0.1)	−0.22(0.09)
$\beta 4$	−0.22(0.06)	−0.3(0.08)	−0.26(0.08)	−0.21(0.06)	−0.30(0.08)	−0.26(0.09)
$\beta 5$	−0.24(0.08)	−0.23(0.1)	−0.25(0.1)	−0.25(0.1)	−0.23(0.11)	−0.26(0.1)
$\beta 6$	−0.21(0.07)	−0.27(0.1)	−0.27(0.09)	−0.20(0.07)	−0.26(0.09)	−0.26(0.09)
deviance	2593(15.8)	2622(16.2)	662(16.4)	2592 (15.4)	2622(15)	660(16)
	$(a = 0.001, b = 0.004)$			$(a = 0.004, b = 0.001)$		
$\beta 1$	7.38(0.09)	7.45(0.12)	7.40(0.1)	7.39(0.09)	7.47(0.11)	7.38(0.11)
$\beta 2$	−0.26(0.09)	−0.28(0.11)	−0.27(0.1)	−0.25(0.09)	−0.27(0.12)	−0.27(0.1)
$\beta 3$	−0.18(0.07)	−0.23(0.1)	−0.22(0.09)	−0.18(0.07)	−0.23(0.1)	−0.22(0.09)
$\beta 4$	−0.22(0.06)	−0.30(0.9)	−0.26(0.08)	−0.22(0.06)	−0.30(0.09)	−0.26(0.08)
$\beta 5$	−0.24(0.09)	−0.23(0.11)	−0.26(0.1)	−0.26(0.09)	−0.25(0.11)	−0.25(0.1)
$\beta 6$	−0.21(0.07)	−0.26(0.1)	−0.26(0.08)	−0.21(0.07)	−0.26(0.1)	−0.26(0.1)
deviance	2593(16)	2622(15.9)	663(16.5)	2594(15.9)	2625(16)	660(16)
	$(a = 0.001, b = 0.005)$			$(a = 0.005, b = 0.001)$		
$\beta 1$	7.37(0.09)	7.45(0.12)	7.38(0.11)	7.36(0.09)	7.46(0.12)	7.38(0.13)
$\beta 2$	−0.27(0.09)	−0.27(0.11)	−0.27(0.11)	−0.26(0.09)	−0.28(0.11)	−0.28(0.11)
$\beta 3$	−0.18(0.08)	−0.23(0.1)	−0.22(0.09)	−0.18(0.07)	−0.22(0.1)	−0.22(0.07)
$\beta 4$	−0.21(0.06)	−0.30(0.09)	−0.26(0.08)	−0.21(0.06)	−0.3(0.09)	−0.26(0.1)
$\beta 5$	−0.23(0.09)	−0.23(0.1)	−0.25(0.1)	−0.23(0.09)	−0.24(0.11)	−0.25(0.13)
$\beta 6$	−0.20(0.07)	−0.26(0.09)	−0.26(0.09)	−0.21(0.07)	−0.26(0.09)	−0.26(0.07)
deviance	2592(15.8)	2622(15.4)	660(16.3)	2591(15.7)	2623(16)	660(15.9)

5.20 Summary

In this chapter, we describe the development of the R package 'afthd', which provides user-friendly implementations for the Accelerated Failure Time (AFT) model. The package includes various specifications for analyzing survival data with missing values, competing risk, selection of appropriate models, and data with augmented entries. Furthermore, we propose a combined technique for selecting covariates and applying the AFT model with Bayesian

TABLE 5.5: Univariate estimates of selected covariate by LASSO with conventional AFT model for 1st

head(rglaft(4,1884, STime='os', Event='death', alpha=1, data=dt1))				
Biomarkers	**Estimate**	**Std. Error**	**Z value**	**Pr(z)**
hsa.mir.149	−0.372	0.067	−5.529	3.213e-08
hsa.mir.7.3	−0.895	0.165	−5.421	5.934e-08
hsa.mir.6728	−1.118	0.214	−5.219	1.80e-07
hsa.mir.4792	−3.848	0.769	−5.001	5.699e-07
hsa.mir.212	−0.543	0.110	−4.915	8.894e-07
hsa.mir.4771.2	−2.420	0.521	−4.648	3.353e-06

TABLE 5.6: Univariate Bayesian estimates using AFT model for selected covariate by LASSO (used previous function *rglaft* for selection)

wbysuni(m=3, n=7, STime='os', Event='death', 3, 10000, data=d22)								
hsa.mir	**coef**	**SD**	**2.50%**	**25%**	**50%**	**75%**	**97.50%**	**Dev**
149	−0.498	0.089	−0.675	−0.556	−0.498	−0.438	−0.324	2617
7.3	−0.372	0.07	−0.504	−0.42	−0.372	−0.325	−0.231	2613
4792	−0.313	0.066	−0.438	−0.358	−0.317	−0.269	−0.176	2616
212	−0.447	0.092	−0.634	−0.506	−0.447	−0.384	−0.269	2611
4771.2	−0.33	0.07	−0.467	−0.377	−0.33	−0.282	−0.192	2617

TABLE 5.7: Multivariable Bayesian estimates using AFT model for selected covariate by LASSO (used *rglaft* function for selection)

wbysmv(m=3, n=7, STime='os', Event='death', 3,10000, data=d22)									
	mean	**sd**	**2.50%**	**25%**	**50%**	**75%**	**97.5%**	**Rhat**	**n.eff**
α	0.962	0.062	0.84	0.92	0.961	1.004	1.087	1.004	620
$\beta 1$	7.383	0.0903	7.216	7.32	7.378	7.441	7.578	1.001	3000
$\beta 2$	−0.257	0.096	−0.448	−0.319	−0.256	−0.194	−0.067	1.002	1500
$\beta 3$	−0.182	0.075	−0.327	−0.233	−0.181	−0.132	−0.033	1.002	1200
$\beta 4$	−0.218	0.0626	−0.34	−0.26	−0.22	−0.178	−0.091	1.002	1600
$\beta 5$	−0.241	0.098	−0.431	−0.307	−0.242	−0.174	−0.043	1.001	3000
$\beta 6$	−0.209	0.073	−0.352	−0.256	−0.212	−0.162	−0.056	1.001	2800
Dev	2593	16	2564	2582	2593	2604	2626	1.001	3000
σ	1.044	0.068	0.92	0.996	1.04	1.087	1.19	1.004	620
τ	0.93	0.121	0.706	0.846	0.924	1.008	1.182	1.004	620

setting in analyzing high-dimensional data. The proposed method fills the gap in analyzing high-dimensional data with the AFT model using Bayesian. The combined technique allows selecting covariates and applying the AFT model with Weibull, log-normal, and log-logistic distribution, with inference based

TABLE 5.8: Univariate Bayesian estimates using conventional AFT model for selected covariate by LASSO for 2nd data

head(rglaft(14,18539, STime= 'os', Event = 'death', 1, dt2))				
	Estimate	Std.Error	z.value	Pr.z.
ILMN_1794643	0.001	0.000	4.959	7.10E-07
ILMN_1699496	−0.000	0.000	−4.958	7.12E-07
ILMN_1795183	−0.000	0.000	−4.849	1.24E-06
ILMN_1693598	−0.002	0.000	−4.809	1.52E-06
ILMN_2070052	−0.000	0.000	−4.725	2.30E-06
ILMN_2388263	−0.010	0.002	−4.719	2.37E-06

TABLE 5.9: Multivariable Bayesian estimates using AFT model for selected covariates by LASSO for 2nd data

wbysmv(m=3, n=7, STime='os', Event='death', 2, 10000, data=dd2)									
	mean	sd	2.50%	25%	50%	75%	97.5%	Rhat	n.eff
α	1.998	0.215	1.597	1.851	1.992	2.142	2.438	1.001	2000
$\beta1$	8.721	0.085	8.57	8.663	8.712	8.772	8.907	1.002	930
$\beta2$	0.222	0.065	0.1	0.175	0.22	0.266	0.35	1.003	1500
$\beta3$	−0.177	0.055	−0.288	−0.213	−0.175	−0.139	−0.075	1.001	2000
$\beta4$	−0.13	0.065	−0.262	−0.174	−0.131	−0.086	−0.004	1.002	1200
$\beta5$	−0.099	0.05	−0.194	−0.133	−0.1	−0.068	0.001	1.001	2000
$\beta6$	−0.175	0.062	−0.295	−0.217	−0.174	−0.134	−0.044	1.002	1000
Dev	12034	11	1184	1196	1203	1211	1227	1.002	1300
σ	0.506	0.056	0.41	0.467	0.502	0.54	0.626	1.001	2000
τ	4.038	0.864	2.552	3.425	3.97	4.589	5.942	1.001	2000

on Markov Chain Monte Carlo (MCMC) simulation. This formulation enables univariate and multivariate analysis of high-dimensional data and identifies the set of imperative regression parameters, their credible intervals, and estimated survival time.

The 'afthd' package is based on a parametric AFT model for Bayesian and conventional interface on high-dimensional settings for multivariate and univariate estimates. The package offers functions that provide estimates of parameters for different distributions in the parametric AFT model and estimates for the conventional approach of AFT model with smooth time functions associated with variable selection using regularization. Moreover, the package provides step-by-step guidelines to use its functions on a high-dimensional dataset, such as the head and neck cancer data, available in the package, to fit all possible models.

The 'afthd' package is capable of working with missing covariates by imputation, and two functions examine data with competing risk models. Additionally, the package has a function that supports augmented data and enables analysis with multivariate. One of the package's advantages is that it requires minimal coding effort to monitor changes in data when there are missing values in covariates, making it easy to use. In summary, the afthd package provides a powerful tool for analyzing survival data with missing values, competing risk, selection of appropriate models, and data with augmented entries. Furthermore, it fills the gap in analyzing high-dimensional data with the AFT model using Bayesian. The package's features allow for the identification of the set of imperative regression parameters, their credible intervals, and estimated survival time in univariate and multivariate analysis of high-dimensional data. The afthd package is easy to use and requires minimal coding effort, making it a valuable addition to the existing R packages for survival analysis.

5.21   Accelerated Failure Time Using R

Accelerated Failure Time (AFT) modeling is a method for modeling time-to-event data, where the focus is on modeling the scale or the distribution of the time-to-event variable, rather than the probability of the event occurring. AFT models assume that the time-to-event follows a particular distribution and that the scale parameter of that distribution depends on one or more covariates.

The most commonly used distributions in AFT models are the exponential and Weibull distributions. AFT models can be used in a wide range of applications, including survival analysis, reliability analysis, and event history analysis.

In R, there are several packages that can be used to perform AFT modeling, such as "survival", "flexsurv", and "aftgee".

The "survival" package provides the function "survreg()" which can be used to fit AFT models with the exponential and Weibull distributions. The function provides estimates of the parameters of the distribution, as well as the log-likelihood and residuals.

The "flexsurv" package provides a more flexible approach to AFT modeling. The "flexsurvreg()" function can be used to fit a wide range of AFT distributions, including the exponential, Weibull, and more complex distributions such as the lognormal and loglogistic distributions.

The "aftgee" package provides a function "aftgee()" which can be used to fit AFT models with complex dependencies between the covariates and the failure time. The function provides estimates of the parameters of the distribution, as well as the log-likelihood and residuals.

It's important to note that AFT models require specifying a distributional assumption, and it's recommended to consult with experts in survival analysis to ensure that the assumptions are appropriate for the data and the research question.

Chapter 6

Longitudinal Modeling

6.1 Introduction

Longitudinal modeling is a statistical method used to analyze data collected over time, also called "panel data", "repeated measures", or "time series" data. The goal of longitudinal modeling is to understand how a response variable changes over time in relation to one or more predictor variables. Longitudinal data can be continuous, categorical, or a combination of both.

There are several different types of longitudinal models, such as:

Mixed-effects models: These models account for both fixed and random effects, allowing for the modeling of both population-level and individual-level effects. Growth curve models: These models are used to model the shape of the growth trajectory over time. Multilevel models: These models account for the hierarchical or nested structure of the data, such as when observations are collected at multiple levels, such as individuals within groups. Longitudinal modeling is widely used in fields such as medicine, psychology, sociology, and economics to study the changes in a response variable over time, and to identify factors that influence these changes. They are particularly useful when there is a need to understand the dynamics of the change of a variable over time.

6.2 Mixed-Effects Regression Models for Continuous Outcomes

Mixed-effects regression models for continuous outcomes, also known as mixed-effects linear models, are a class of statistical models that are used to analyze data collected over time, also called "repeated measures" data. These models are used to estimate the relationship between a continuous response variable and one or more predictor variables, while accounting for both fixed and random effects.

The fixed effects in the model are the predictors that are believed to have an effect on the response variable, such as time or treatment. The random effects in the model are the subject-specific effects, such as the

DOI: 10.1201/9781003298373-6

individual-specific intercepts or slopes, which are assumed to be random variables with a probability distribution. The random effects account for the dependence within the data, such as the dependence between repeated measures for the same individual.

Mixed-effects regression models for continuous outcomes are widely used in fields such as medicine, psychology, and engineering to analyze data collected over time. They provide a powerful tool for estimating population-level effects while accounting for individual-level differences and dependencies. They are also useful when the goal of the analysis is to understand how a response variable changes over time in relation to one or more predictor variables.

6.3   Mixed-Effects Polynomial Regression Models

Mixed-effects polynomial regression models are a class of statistical models that extend mixed-effects linear models by allowing the relationship between the response variable and the predictor variables to be non-linear. These models are used to analyze data collected over time, also called "repeated measures" data. They allow the modeling of a more complex relationship between the response variable and the predictor variables, while accounting for both fixed and random effects.

In mixed-effects polynomial regression models, the predictor variables are transformed into polynomial functions of different degrees. This allows for a more flexible modeling of the relationship between the response variable and the predictor variables. The fixed effects in the model are the polynomial terms, such as linear, quadratic or cubic, that are believed to have an effect on the response variable, such as time or treatment. The random effects in the model are the subject-specific effects, such as the individual-specific intercepts or slopes, which are assumed to be random variables with a probability distribution.

Mixed-effects polynomial regression models are useful when the relationship between the response variable and the predictor variables is non-linear and when the goal of the analysis is to understand how a response variable changes over time in relation to one or more predictor variables. They are widely used in fields such as medicine, psychology, engineering, and economics to analyze data collected over time.

6.4   Covariance Pattern Models

Covariance pattern models are a class of statistical models that are used to analyze multivariate data, particularly in the context of longitudinal data or clustered data. The goal of these models is to understand the relationship between multiple response variables and one or more predictor variables, while accounting for the dependence structure among the response variables.

There are several types of covariance pattern models, such as:

Structural equation modeling (SEM): These models are used to estimate the relationships among multiple observed and latent variables, and to model the covariance structure of the data. Latent class analysis (LCA): These models are used to identify subgroups of individuals based on their patterns of response variables and to estimate the relationships between the subgroups and predictor variables. Latent profile analysis (LPA): These models are used to identify subgroups of individuals based on their patterns of response variables and to estimate the relationships between the subgroups and predictor variables. Covariance pattern models are particularly useful when the goal of the analysis is to understand the relationship between multiple response variables and one or more predictor variables, while accounting for the dependence structure among the response variables. They are widely used in fields such as psychology, sociology, and epidemiology to analyze multivariate data, particularly

6.5 Mixed Regression Models with Autocorrelated Errors

Mixed regression models with autocorrelated errors are a class of statistical models that are used to analyze data collected over time, also known as "panel data" or "repeated measures" data. These models are used to estimate the relationship between a response variable and one or more predictor variables, while accounting for both fixed and random effects and for the presence of autocorrelation in the errors.

Autocorrelation refers to the presence of correlation between errors at different time points. In mixed regression models with autocorrelated errors, the errors are assumed to be correlated over time, and a specific structure is assumed for this correlation. Common structures include an autoregressive (AR) structure, where the error at time t is correlated with the error at time t–1 and a moving average (MA) structure, where the error at time t is correlated with the average of the errors over a certain period of time.

These models are useful when the goal of the analysis is to understand the relationship between a response variable and one or more predictor variables over time, while accounting for both fixed and random effects and for the presence of autocorrelation in the errors. They are widely used in fields such as finance, economics, and engineering to analyze data collected over time.

6.6　Generalized Estimating Equations (GEE) Models

GEE models are a class of statistical models that are used to analyze data collected over time, also known as "panel data" or "repeated measures" data. These models are used to estimate the relationship between a response variable and one or more predictor variables, while accounting for the dependence structure among the observations.

GEE models are an extension of Generalized Linear Models (GLM), which allows for the modeling of correlated data by specifying a working correlation matrix, which describes the dependence structure among the observations. GEE models estimate the population-average parameters and allow for the modeling of both within-subject and between-subject effects.

GEE models are useful when the goal of the analysis is to understand the relationship between a response variable and one or more predictor variables over time, while accounting for the dependence structure among the observations. They are widely used in fields such as medicine, psychology, sociology, and epidemiology to analyze data collected over time. They are particularly useful when the data has a clustered structure, such as multiple observations for each individual, or when the data is not independent and identically distributed.

In R, the package geepack provides the geese() function which can be used to fit GEE models. The function takes a formula as input and the model should include the response variable as the left-hand side of the formula and the predictor variables as the right-hand side of the formula. Additionally, the geese() function allows the user to specify the correlation structure of the data, such as an exchangeable, independent, or autoregressive structure.

For example, if the data is in a dataframe called 'data' and the response variable is called 'y', the predictor variable is called 'x' and the subject variable is called 'id', the GEE model can be fit using the following code:

```
library(geepack)
model=geese(y ~ x, id = id, corstr = "exchangeable", data = data)
```

The package gee also provides similar functionality, but with a more general implementation, it has many options for specifying the correlation structure and the family distribution of the response variable.

Once the model is fit, the summary() function can be used to obtain the estimates of the parameters, standard errors, and p-values, and the predict() function can be used to predict the response variable for new data.

6.7   Missing Data Analysis

Missing data analysis is a statistical method used to handle missing data in a dataset. Missing data can occur for a variety of reasons, such as nonresponse, measurement error, or data entry errors. The goal of missing data analysis is to minimize the bias and loss of precision that can occur when data is missing, and to make valid inferences based on the available data.

There are several different methods for handling missing data, including:

Complete case analysis (or listwise deletion): This method involves removing all observations that have any missing data. This approach is simple but can result in bias and loss of precision if the data is missing at random (MAR) or missing completely at random (MCAR). Mean imputation: This method involves replacing missing values with the mean of the observed values. This can be useful when the data is missing at random, but can introduce bias if the data is missing not at random (MNAR). Multiple imputation: This method involves generating multiple imputed datasets, and then analyzing each imputed dataset separately. The results are then combined to account for the uncertainty introduced by the missing data. This method can provide valid inferences under MAR and MNAR assumptions. It is important to note that missing data analysis is not a standalone method, it is a step that should be integrated in the overall research design and data analysis plan. The best approach to missing data depends on the assumptions about how the data is missing and the specific research question. Missing data analysis is a type of statistical analysis that is used to handle missing values in a dataset. In R, there are several packages that can be used to perform missing data analysis, such as the mice package, the Amelia package, and the MissMech package.

One of the most commonly used packages for missing data analysis in R is the mice package which provides multiple imputation methods. The package provides the mice() function, which can be used to impute missing values in a dataset. The function takes a dataframe as input and creates multiple imputed datasets by using a specified method (such as predictive mean matching or fully conditional specification) to fill in the missing values.

For example, if the data is in a dataframe called 'data', the imputed datasets can be created using the following code:

```
library(mice)
imputedData=mice(data)
```

The Amelia package provides similar functionality to the mice package, but it also allows for the handling of missing data in time series data. The package provides the amelia() function which can be used to impute missing values in a dataset. The function also takes a dataframe as input and creates multiple imputed datasets by using a specified method (such as predictive mean matching or fully conditional specification) to fill in the missing values.

For example, if the data is in a dataframe called 'data', the imputed datasets can be created using the following code:

```
library(Amelia)
imputedData=amelia(data)
```

The MissMech package provides a different approach to missing data analysis, it provides the missMech() function which can be used to fit a missing data mechanism model. It estimates the probability of missing data for each variable and uses this information to impute missing values.

Once the imputed datasets are created, they can be used for further analysis. It is important to note that when working with imputed data, it is important to use appropriate methods for estimating standard errors and p-values, such as the Rubin's rules, to account for the uncertainty introduced by the imputation process.

It is important to note that, the choice of imputation method will depend on the type of data, the pattern of missing data, and the research question. It is also important to check the assumptions of the imputation method and the quality of the imputed data using graphical methods and statistical tests before interpreting the results.

6.8   Multiple Imputation

Multiple imputation is a statistical method used to handle missing data in longitudinal studies. Longitudinal studies are studies that collect data on the same individuals over time, often with repeated measures of the outcome variable. These studies are prone to missing data, which can lead to biased estimates of the parameters of interest if not handled properly.

Multiple imputation is a way to handle missing data by creating multiple imputed datasets, each with different, but plausible, values for the missing data. The idea is that each imputed dataset reflects the uncertainty about the true values of the missing data. By analyzing each imputed dataset separately and then combining the results, multiple imputation provides more accurate and reliable estimates of the parameters of interest.

In R, there are several packages that can be used to perform multiple imputation for longitudinal data, such as "mice", "Amelia", and "mi".

The "mice" package provides a comprehensive framework for multiple imputation of missing data, including functions for imputing missing data, analyzing the imputed data, and combining the results. This package can handle longitudinal data with a variety of missing data mechanisms, such as missing at random and missing not at random.

The "Amelia" package provides a function "amelia()" which can be used to perform multiple imputation for longitudinal data. This function can handle longitudinal data with a variety of missing data mechanisms, and can also handle complex data structures such as multilevel or clustered data.

The "mi" package provides an easy and simple way for multiple imputation for longitudinal data and it provides a function "mi()" which can be used to perform multiple imputation for longitudinal data, it can handle various missing data mechanisms, and it can also handle complex data structures such as multilevel or clustered data.

It's important to note that multiple imputation is a complex method and it's recommended to consult with experts in missing data analysis to ensure that the analysis is correctly performed.

6.9 Multiple Imputation Using Mediation Analysis

Multiple imputation can also be used in combination with mediation analysis when dealing with missing data in longitudinal studies. Mediation analysis is a method used to identify the mechanisms or pathways through which an exposure affects an outcome, and it can be used in conjunction with longitudinal data to understand how the mechanisms change over time.

When performing mediation analysis with longitudinal data, missing data can be a problem as it can lead to biased estimates of the parameters of interest. Multiple imputation can be used to handle missing data by creating multiple imputed datasets, each with different but plausible values for the missing data.

In R, there are several packages that can be used to perform multiple imputation in longitudinal data mediation analysis, such as "mediation", "medflex", and "causMed".

The "mediation" package provides a function "mediate()" which can be used to perform mediation analysis with multiple imputed datasets. The function can handle both continuous and categorical mediators and outcomes, and can also handle multiple imputed datasets created with the "mice" package.

The "medflex" package provides a function "medflex()" which can be used to perform mediation analysis with multiple imputed datasets. It can handle both continuous and categorical mediators and outcomes and can also handle complex data structures such as multilevel or clustered data.

The "causMed" package also provides a function "mediation()" which can be used to perform mediation analysis with multiple imputed datasets. It can handle both continuous and categorical mediators and outcomes and can also handle complex data structures such as multilevel or clustered data and it's based on the Bayesian framework.

It's important to note that multiple imputation in longitudinal data mediation analysis is a complex method and it's recommended to consult with experts in causal inference and missing data analysis to ensure that the analysis is correctly performed.

6.10   Longitudinal Data Analysis Using R

Longitudinal data analysis is a type of statistical analysis that is used to study changes in a response variable over time. In R, there are several packages that can be used to perform longitudinal data analysis, such as the lme4 package, the nlme package, and the mixed package.

One of the most commonly used packages for longitudinal data analysis in R is the lme4 package which provides the lmer() function. This function can be used to fit linear mixed-effects models (LMM) and can handle both fixed effects, which are estimated for all observations, and random effects, which are specific to individual subjects. The lmer() function takes a formula as input and the model should include the response variable as the left-hand side of the formula, the time variable as a predictor, and any other predictor variables as the right-hand side of the formula.

For example, if the data is in a dataframe called 'data' and the response variable is called 'y', the time variable is called 'time' and the subject variable is called 'id', the linear mixed-effects model can be fit using the following code:

```
library(lme4)
model=lmer(y~ time + (time—id), data = data)
```

The nlme package provides similar functionality to the lme4 package and also provides the nlme() function which can be used to fit linear and non-linear mixed-effects models. The nlme() function also takes a formula as input and the model should include the response variable as the left-hand side of the formula, the time variable as a predictor, and any other predictor variables as the right-hand side of the formula.

For example, if the data is in a dataframe called 'data' and the response variable is called 'y', the time variable is called 'time' and the subject variable is called 'id', the non-linear mixed-effects model can be fit using the following code:

```
library(nlme)
model=nlme(y ~ SSasymp(time,Asym,R0),
fixed=Asym+R0~    1,random=pdDiag(list(Asym~    1,    R0    ~
1)),data=data,group=id)
```

The mixed package also provides similar functionality, but it is more flexible in handling multiple random effects and it also allows for correlation structure in the random effects.

Once the model is fit, the summary() function can be used to obtain the estimates of the parameters, standard errors, and p-values, and the predict() function can be used to predict the response variable for new data.

It is important to note that when working with longitudinal data, it is important to check the assumptions of the model and the goodness-of-fit of the model using graphical methods and statistical tests before interpreting the results.

Bayesian mixed effect model with MCMC

```
data(repdata)
Bysmixed(m=4,n=7,t="Age",group="Gender",chains=4,n.adapt=100,
repdata)
```

Bayesian mixed effect model for high dimensional longitduinal data with deviance information criterion (DIC)

```
data(msrep)
BysmxDIC(m=c(4,8,12),tmax=4,t="Age",
group="Gender",chains=4,iter=1000,out="DIC",data=msrep)
```

Bayesian mixed effect model for high dimensional longitduinal data with highest posterior density. interval

```
data(msrep)
BysmxHPD(m=c(4,8,12),tmax=4,t="Age",
group="Gender",chains=4,iter=1000,out="hpD",data=msrep)
```

Uses bonferroni correction factor in survival analysis

```
data(mesrep)
Bysmxms(m=4,n=7,time="Age",
group="Gender",chains=4,n.adapt=100,data=msrep)
```

TABLE 6.1:   Missing not at random by MCMC

Parameter	Mean	SD	2.5%	97.5%	Rhat
diff	9.085	15.360	−15.059	29.104	1.113
mean1[1]	68.978	9.254	59.361	85.227	1.020
mean1[2]	62.572	10.499	43.313	75.655	1.070
mean2[1]	79.726	10.141	63.262	94.192	0.992
mean2[2]	85.340	10.463	66.520	97.779	1.430
mean3[1]	68.390	8.838	55.986	80.840	1.051
mean3[2]	77.475	12.339	56.160	90.922	0.973
prob0	0.700	0.483	0.000	1.000	0.989
sigma1[1]	34.190	6.556	24.411	44.500	2.879
sigma1[2]	42.135	5.629	35.557	49.840	4.986
sigma2[1]	47.202	11.022	34.349	61.093	8.195
sigma2[2]	37.740	6.086	29.355	47.186	2.644
sigma2_1[1]	15.282	1.367	13.351	17.474	1.067
sigma2_1[2]	31.377	2.282	27.395	34.187	1.111
sigma3[1]	24.359	2.653	21.014	28.494	2.922
sigma3[2]	29.094	7.068	18.381	39.685	3.462
sigma3_2_1[1]	16.743	4.982	9.327	22.071	3.099
sigma3_2_1[2]	19.228	2.735	15.411	22.703	2.209
deviance	1678.832	21.831	1695	1701.715	3.002

Bayesian mixed model with random intercepts and random slopes for high dimensional longitudinal data with batch size

```
data(hnscc)
hidimSurvbonlas(6,104,0.05,ID="id",OS="os",
Death="death",PFS="pfs",Prog="prog",hnscc)
```

Bayesian multivariate regression with unstructured covariance matrix for high dimensional longitudinal data

```
data(repdata)
creg(m=4,n=7,chains=4,n.adapt=100,data=repdata)
```

Creating batches of variables on high dimensional data

```
data(gh)
hdmarjg(m=1,n=3,treatment="Treatment",
n.chains=2,n.iter=10,dat=gh)
```

TABLE 6.2: Bayesian multivariate normal regression with unstructured covariance matrix for high-dimensional longitudinal data

Parameter	mu(sd)	[2.5%,97.5%]	Rhat	n.eff
Omega1[1,1]	0.42(0.10)	[0.27,0.64]	0.99	200
Omega1[2,1]	−0.16(0.09)	[−0.38,−0.00]	0.99	200
Omega1[3,1]	−0.22(0.08)	[−0.39,−0.07]	1.00	200
Omega1[4,1]	0.05(0.08)	[−0.10,0.19]	1.00	190
Omega1[1,2]	−0.16(0.09)	[−0.38,−0.00]	0.99	200
Sigma1[1,3]	5.01(1.64)	[2.61,8.61]	1.00	200
Sigma1[2,3]	4.05(1.36)	[1.79,7.35]	1.02	120
Sigma1[3,3]	8.06(2.05)	[4.73,12.34]	1.01	180
Sigma1[4,3]	6.40(1.98)	[3.33,10.83]	1.00	200
Sigma1[1,4]	4.15(1.62)	[1.63,7.62]	1.00	200
Sigma1[2,4]	4.74(1.54)	[2.43,8.64]	1.02	94
Sigma1[3,4]	6.40(1.98)	[3.33,10.83]	1.00	200
Sigma1[4,4]	8.11(2.40)	[4.54,13.75]	1.02	99
mu1[1]	21.94(0.34)	[21.24,22.56]	1.01	170
mu1[2]	23.29(0.37)	[22.55,24.01]	1.01	130
mu1[3]	24.64(0.44)	[23.84,25.51]	1.01	130
mu1[4]	25.99(0.53)	[25.07,27.07]	1.01	130
mu2[1]	21.91(0.33)	[21.23,22.48]	0.99	200
mu2[2]	23.25(0.36)	[22.47,23.92]	1.00	200
mu2[3]	24.59(0.43)	[23.71,25.40]	1.00	200
mu2[4]	25.93(0.52)	[24.94,26.93]	1.00	200
Deviance	886.21(6.52)	[876,901]	1.00	200

Missing not at random by MCMC

```
data(gh)
hdmnarjg(m=1,n=3,treatment="Treatment",
n.chains=2,n.iter=10,dat=gh)
```

Bayesian multivariate normal regression with unstructured covariance matrix for high dimensional longitudinal data

```
data(repdata)
mvncovar2(m=4,n=7,time="Age",
group="Gender",chains=4,iter=100,data=repdata)
```

Chapter 7

High-Dimensional Data Analysis

7.1 Introduction

Feature selection is a process of creating a subset of candidate features by dimensional reduction technique, removing irrelevant features, enhancing learning accuracy, and improving result comprehensibility [101]. An appropriate selection of the features can guide toward the improvement of the inductive learner in terms of faster response, minimum generalization error, simplicity of the induced model, and hopefully a better understanding of the domain [102]. This strategy provides us an effective and efficient way to work with high-dimensional data along with machine-learning problems [103]. The task is to develop the procedure from gene expression data and develop a strategy by choosing more relevant genes involved in the progression of a disease. The difficulty lies in the fact that the high-dimensional datasets generally consist of a limited number of replicates, and the sample size is small [104]. Therefore, we need a criterion that uses the information to undergo a feature selection process before building any prediction model. In literature, different feature selection methods usually identify the significant genes or proteins associated with the response variable. However, false-positive genes are sometimes selected in the process because of repeated number of tests and selection bias [105, 106]. A false-positive selection is an error that incorrectly determines a feature is true and it may also happen for performing the reiteration in a process. Selection bias is an experimental error that occurred due to inappropriate selection of sample genes that does not accurately reflect the target population [107]. It occurs as a result of factors that influence subjects' participation continuously in a study. In addition, an improper feature selection procedure also leads to biases. The consequence of the biases is that the association between exposure and outcome is different for those selected in the analysis from the association among those eligible.

We can classify the feature selection methods into three categories: (i) filter methods, (ii) wrapper methods, and (iii) embedded methods [108]. The filter methods select features based on a performance measure and rank individual features or evaluate an entire subset of features regardless of the employed data modeling algorithm. Typical measurements for feature filtering are correlation-based feature selection, analysis of variance, chi-square,

DOI: 10.1201/9781003298373-7

Fisher score, and multi-cluster feature selection. These measurements are generally used to assess the relationships among the response and independent variables [108, 109]. On the contrary, the feature subsets are chosen by the performance quality of a modeling algorithm in the wrapper methods. As a black box evaluator, wrappers evaluate subsets based on the classifier performance for classification and a clustering algorithm for clustering [108, 110]. Forward selection and backward elimination are two standard processes under the wrapper methods. The former process gradually incorporates the variables into larger subsets, whereas the latter one starts with all the variables taken together and progressively eliminates the least promising variables [110]. However, filter methods fail to incorporate learning algorithms, and wrapper methods utilize a learning machine to measure the quality of subsets of variables without incorporating knowledge about the specific structure of the function. They can therefore be combined with any learning machine [111]. The embedded methods cannot separate the learning and feature selection parts and use filter and wrapper methods to choose the independent features. Lasso, Ridge, and Elastic net regularization are some of the popular embedded methods [112, 113].

Simplicity is the main advantage of filter methods, and they do not require any high computational capabilities or a long time to run the process. As opposed to filter methods, wrapper methods fit the data better by selecting relevant features for modeling rather than focusing on the association [114]. The kernel-based approach is a new way of feature selection that assists in reducing the overall dimension of the dataset using some kernel functions [115]. Another existing algorithm on feature selection is recursive feature elimination. This elimination procedure uses the list of features to train the dataset and assign weights for each of the features. The features having the least weight are then removed using a training dataset. Sometimes, we make the assumptions of distribution and select the training parts from the actual data by random bootstrap sampling procedure using those assumptions [116].

The common drawbacks of the above methods are a false-positive or false-negative selection of features and lack of regeneration [105, 117]. Because of selection bias, sometimes the sampling data may wrongly favor or oppose a few variables. Consequently, those procedures do not appropriately select the study participants due to their presence. The results of the embedded methods also depend entirely on the taste of the training and testing datasets. Therefore, an issue of replication arises for choosing some of the variables only owing to a specific combination for selection and validation groups [118]. The re-sampling feature selection algorithm is one of the critical embedded methods that rely on sorting the important features from the entire data using re-sampling techniques. These techniques are dependent on the weight function for sorting the features. [119] defined the weight function as the proportion of each of the features in the entire iteration process. We call it as P_{weight}.

The re-sampling algorithm is adopted to overcome the selection bias because of the taste of training and validated dataset and provide an ordered list of relevant factors according to their weights. The objective of our study is to construct a weight function that helps to select the features based on their consistency and the re-sampling method. We propose a modified version of the weight function to incorporate the concept of consistency in the feature selection method by the coefficient of variation (CV).

7.2   Survival Analysis in High-Dimensional Data

Survival analysis in high-dimensional data is a statistical method used to analyze time-to-event data in the presence of a large number of covariates. High-dimensional data refers to situations where the number of covariates is much larger than the number of observations. The goal of survival analysis in high-dimensional data is to identify the covariates that are associated with the risk of an event occurring, while adjusting for the high-dimensionality of the data.

There are several methods for survival analysis in high-dimensional data, such as:

Lasso: This method uses L1 regularization to select a subset of covariates by shrinking the coefficient estimates toward zero. Elastic Net: This method uses a combination of L1 and L2 regularization to select a subset of covariates. Random Survival Forest: This method uses a random forest algorithm to identify the covariates that are associated with the risk of an event occurring. These methods are particularly useful when the goal of the analysis is to identify the covariates that are associated with the risk of an event occurring, while adjusting for the high-dimensionality of the data. They are widely used in fields such as medicine, epidemiology, and engineering to analyze time-to-event data in high-dimensional settings. Let us consider a time to event dataset containing n individuals and m normalized features or expression values. Also, let the list of features is denoted by $X = \{x_1, x_2, ..., x_j, ..., x_m\}$. We are interested in understanding the effect of the covariates or features on the hazard rate.

The hazard rate is the risk of suffering the event of interest given that the individual has survived up to a specific time, where the event is represented by $\{0, 1\}$ such as "0" means 'no event occurred' and "1" represents 'event occurred'. We use the Cox Proportional Hazards Model to illustrate the algorithm. The proportional hazards model is usually written as

$$\lambda(t \mid x_1, x_2, ..., x_m) = \lambda_0(t)\, e^{x_1\beta_1 + x_2\beta_2 + ... + x_m\beta_m} = \lambda_0(t)\, e^{x^T\beta} \qquad (7.1)$$

where, t is the time of an event, $\lambda_0(t)$ is known as the baseline hazard function, $\beta = \{\beta_1, \beta_2, ..., \beta_j, ..., \beta_m\}$ is the vector of parameters, and $X = \{x_1, x_2, ..., x_j, ..., x_m\}$ is a set of normalized continuous predictors.

We use the partial likelihood function for the estimation of the parameters. Assuming the sample consists of n independent subjects and x_{ij} represents the value of X data matrix for the ith patient corresponding to the jth covariate or normalized predictor, $i = 1, 2, .., n$ and $j = 1, 2, ..., m$. The form of the partial likelihood is generally represented by $PL(\beta)$ and defined as,

$$PL(\beta) = \prod_{l=1}^{d} \left[\frac{exp(\boldsymbol{x}_{i(t_l)}^T \boldsymbol{\beta})}{\sum_{i=1}^{n} exp(\boldsymbol{x}_i^T \boldsymbol{\beta}) Y_i(t_l)} \right] \tag{7.2}$$

where, $t_1, t_2, ..., t_d$ are the distinct event times; $Y_i(t_l)$ denotes the indicator for whether or not the ith patient is at risk at time t_l, then $Y_i(t_l) = 1$ if the patient i is at risk at t_l, otherwise zero; $Y(t_l) = \sum_{i=1}^{n} Y_i(t_l)$ be the total number of entire sample who are at risk at time t_l; and $\boldsymbol{x}_{i(t_l)}^T$ be the vector of covariates corresponding to the patient i at time t_l.

In this illustration, we take the logarithm of the partial likelihood function. Then, the unknown parameters are estimated by maximizing the partial log-likelihood w.r.to $\boldsymbol{\beta}$.

The re-sampling feature selection algorithm begins with selecting the features analogous to the filter methods. When the number of features is superior to the whole sample size, removal the unworthy parts from the data is required. This preliminary screening can be performed through any filter method. Since the illustration is treated for prediction purposes as well as the selection of relevant features using the weight function approach, one should not be concerned about the false-positive selection of variables at this time of analysis. The primary screening will assist us in eliminating the unnecessary features in the next steps, and after that, the dataset is randomly split into training and testing groups. Let the training set and the validation set be denoted by T_i and V_i, respectively. Also, let S_i is the splitting scenario, and F_i be the set of variables selected using S_i and a suitable algorithm. The process is repeated a fixed number of times (say, N) using different training and testing datasets. It would be adequate to take sufficiently large N for getting better results. For each repetition, the T_i and V_i dataset are randomly chosen, and the resulting variables save to a dataset for further scrutiny.

The modified weight function for the feature x_j is defined as

$$w_j^* = exp[-\frac{1}{2} \frac{\sigma_j/\mu_j}{R} (1 - \frac{n_j}{N})] \tag{7.3}$$

where, $j = 1, 2, ..., m$, and μ_j and σ_j be the respective mean and standard deviation of the jth feature.

Assuming the feature x_j is selected in the model n_j times after repeating the process N times ($n_j \leq N$), then $w_j = \frac{n_j}{N}$ is defined as the proportion of times the variable x_j selected to the model. One of the usage statistics for comparing the degree of variation from one data series to another is the coefficient of variation, even if the means are drastically different from one another. Here we are trying to incorporate the concept of consistency through the CV measure. The coefficient of variation of the feature x_j represents the ratio of the standard deviation to the mean, i.e.,

$$CV_j = \frac{\sigma_j}{\mu_j} \tag{7.4}$$

The difference between maximum and minimum of all the CV values is denoted by R and it is defined as

$$R = max\{\frac{\sigma_j}{\mu_j}; 1 \leq j \leq m\} - min\{\frac{\sigma_j}{\mu_j}; 1 \leq j \leq m\} \tag{7.5}$$

If the CV of a feature is lesser than that of another feature, one can conclude that the former feature is more consistent than the latter one. The CV is then divided by R and multiplied with $(1 - w_j)$. We use the logarithmic transformation $-2log_e(w_j^*)$ for modification of w_j. By using this transformation and equations (7.4) and (7.5), the modification of weight function will be derived as

$$\frac{CV_j}{R}(1 - w_j) = -2log_e(w_j^*) \tag{7.6}$$

$$w_j^* = exp[-\frac{1}{2}\frac{CV_j}{R}(1 - w_j)] = exp[-\frac{1}{2}\cdot\frac{\sigma_j/\mu_j}{R}(1 - \frac{n_j}{N})] \tag{7.7}$$

For fixed values of n_j and N, if the coefficient of variation of x_j is exceeding than others, we consider x_j as a feature with less consistency. As a result, the numerical value of the modified weight function of x_j is always minor compared to a more consistent feature. So, the new weight of a less consistent feature is always inferior to a more consistent one. Mathematically, w_j^* will always lie between 0 and 1. When the product of the coefficient of variation and $(1 - w_j)$ tends to infinity, the modified weight function approaches its value toward zero. This happens only when w_j tends to zero, and the coefficient of variation is large enough. When the CV of x_j goes to zero and n_j takes its value around N, the new weight converges to 1. It is notable that in case the mean is negative, modulus has to be considered to make it positive so that $0 \leq w_j^* \leq 1$, for all j. Suppose we call it by C_{weight}.

The new weight function will help us to select the more consistent features and order them according to their significance in the model. Features with zero weightage are eliminated in the first step. Also, there will be features with less consistency, and smaller weightage dropped off from the model in the other iteration process.

Let us consider an ordered list of features $X^* = \{x_{(1)}, x_{(2)}, ..., x_{(j)}, ..., x_{(m)}\}$ appeared at least once in the selection process such that $x_{(m)}$ gets the maximum weight and $x_{(1)}$ gets the minimum weight. Now having an ordered list of features, we can fix a proportion p such that the feature $x_{(j)}$ will be included in the final model only if the corresponding $w_j^* > p$. The algorithm can be summarized as following steps:

- Step 1. If the number of features is much larger than the number of samples, the filter method is used with the relaxed cut-off values to choose the factors initially.

- Step 2. Split the dataset randomly into training and testing groups and run a wrapper algorithm for selecting the features.

- Step 3. Repeat step 2 for N number of times and store the selected factors into a dataset.

- Step 4. Calculate the new weight for each of the selected factors using equation (7.3).

- Step 5. An ordered list of features is then created according to their modified weights.

- Step 6. Fix a cut-off weight and include those features which have a weight more than the cut-off value to the final model.

After fitting the model, it is always required to evaluate how well the model fits the data it was generated from. The Akaike information criterion (AIC) can help us in this context. In statistics, the AIC is used for comparison among different possible models and to determine which one is the best fit for the data [120]. This mathematical method is calculated based on the number of independent variables used for model building and the maximum likelihood estimate of the model. The AIC formula is:

$$AIC = -2lnL(\hat{\beta} \mid X) + 2K \qquad (7.8)$$

where, $L(\hat{\beta} \mid X)$ represents the likelihood of $\hat{\beta}$ given data X, $\hat{\beta}$ is the maximum likelihood estimate of the parameter vector β given X, and K is the number of estimated parameters. The best-fit model according to AIC is the one which has minimum AIC value.

7.3   Challenges in High-Dimensional Data

High-dimensional data refers to data that has a large number of variables or features, often with a relatively small number of observations. Handling high-dimensional data can be challenging as it may require specialized techniques that are different from those used for low-dimensional data. Some of the challenges of high-dimensional data include:

Curse of dimensionality: As the number of variables or features increases, the amount of data required to accurately estimate the model parameters also increases. This can make it difficult to find patterns or relationships in the data.

Overfitting: With a large number of variables or features, it is more likely that a model will find patterns in the data that are specific to the training set and not generalizable to new data. This is known as overfitting and can lead to poor model performance on new data.

Variable selection: With a large number of variables or features, it can be difficult to determine which variables are important for the analysis. Variable selection techniques such as lasso or Ridge regression can be used to identify a subset of important variables.

Feature engineering: With a large number of variables or features, it can be difficult to identify and create new variables or features that are relevant to the analysis.

Computational challenges: High-dimensional data can require significant computational resources to process and analyze. Specialized algorithms, such as dimensionality reduction techniques, are often required to handle the data.

Visualization: High-dimensional data can be difficult to visualize, as it may require specialized techniques such as parallel coordinates plots, multidimensional scaling, or other visualization techniques to represent the data in a meaningful way.

Overall, high-dimensional data requires a combination of statistical and computational methods to handle it effectively, and addressing these challenges is important for extracting meaningful insights from the data.

The challenges in high-dimensional survival data analysis is performed with available gene expression dataset. The dataset is obtained from Gene Expression Omnibus (GEO) (https://www.ncbi.nlm.nih.gov/geo/) with accession number GSE 65622. The dataset consists of 507 proteins expression values from 80 locally advanced rectal cancer patients with multiple treatment interventions and 5 years follow-up duration. The relapse of the disease is our event of interest. The baseline expression data of 507 proteins is used to show the method. It is verified that a total of 28 (35%) subjects has experienced the event during the study tenure [118, 119]. After downloading the dataset in CSV format, we used MS Excel to clean the dataset preliminary. R version 3.6.3 (Copyright (C) 2020 The R Foundation for Statistical Computing) is used to create the function "weightfun"(https://github.com/soutik28/Modified-Weight-Function/commit/main) and run the algorithm. This function is coupled with 'glmnet' and 'survival' packages in R [121].

All the proteins with 5% missing values were omitted from the data during the initial screening, and we got a total of 327 (64.5%) protein expression values. It might happen that some clinically important proteins were excluded due to the missing values since the data were used only for illustrating the method—proteins' clinical importance can be ignored. The mean of the expression values then replaces the missing values of the chosen proteins to obtain a complete dataset. Since the number of expression values was more than four times higher than the sample size, we have used the univariate Cox-proportional hazards model with a 5% level of significance for further filtering

TABLE 7.1: List of significant proteins based on level of significance 0.05.

Proteins	P-value	Proteins	P-value
Activin	0.042	Activin	0.024
Adiponectin	0.030	AgRP	0.015
Angiogenin	0.000	BCMA	0.004
BMP7	0.032	BTC	0.033
Cardiotrophin	0.012	CCR8	0.020
CCR9	0.001	CD14	0.011
CD27	0.000	Chordin2	0.004
CRTH	0.005	CTACK	0.000
CXCR4	0.000	Decorin	0.004
EGF	0.002	CD105	0.002
Endothelin	0.009	CCL26	0.032
FADD	0.002	FGF13	0.000
FGF 5	0.009	FGF6	0.016
FGF 9	0.024	Follistatin	0.030
Frizzled 4	0.000	Galectin3	0.027
GASP	0.004	GDF11	0.000
GRO	0.000	HCC4	0.000
Hepassocin	0.011	HGF	0.007
HRG	0.030	ICAM1	0.038
ICAM	0.000	IFN	0.003
IFNR2	0.021	IFN	0.001
IGFBP 7	0.030	IL1	0.000
IL1ra	0.025	IL1	0.006
IL10R	0.011	IL-11	0.040
IL12-p70	0.006	IL-13	0.011
IL-17RC	0.000	IL 22 R	0.036
IL28A	0.010	IL-5 R	0.000
IL-7	0.011	Insulysin	0.021
IP10	0.048	Kremen1	0.007
Kremen	0.004	CD62L	0.029

the features. The results provided us with 100 significant proteins in the first step of the algorithm. Table 7.2 shows the set of significant proteins and their corresponding p-values. Using the C_{weight} and a threshold value $p = 0.90$, we selected 12 proteins from a set of 100 to model the dataset for further analysis. These 12 proteins are the most relevant in terms of their consistency and occurrence in the iteration process. The estimated parameters corresponding to these 12 proteins shown in Table 7.3 are the $\hat{\beta}_j\{j = 1, 2, ..., 12\}$ values for the final multivariate Cox-proportional hazards model. Table 7.4 compares the feature selection of the re-sampling method using C_{weight} with the same method using P_{weight}. We get an AIC 120.22 for the model on C_{weight} whereas the AIC value for the model on P_{weight} is 165.45. The predictive model on

TABLE 7.2: List of significant proteins based on level of significance 0.05.

Proteins	P-value	Proteins	P-value
Lck	0.019	Lipocalin 2	0.013
LRP 1	0.003	MCP 3	0.008
MFRP	0.018	MIF	0.000
MIP 3 beta	0.040	MMP1	0.001
MMP 14	0.037	MMP15	0.047
MMP 20	0.000	MSP	0.000
NAP 2	0.005	NCAM1	0.044
NOV CCN3	0.044	OrexinA	0.024
Orexin B	0.009	OSM	0.001
Osteocrin	0.000	OX40	0.030
PD	0.005	PDGF	0.036
PDGF	0.042	PDGF	0.012
PF4	0.000	PLUNC	0.010
SDF1	0.032	Siglec5	0.045
Siglec 9	0.005	SPARC	0.003
Thrombospondin1	0.019	Thrombospondin2	0.009
TIMP2	0.000	TIMP 3	0.026
Tomoregulin1	0.000	TNFRSF1B	0.013
TNF	0.002	TNFRSF10A	0.035
TNFRSF10C	0.013	TRANCE	0.028

TABLE 7.3: Proteins list with their weights and CV after 150 iterations using threshold 0.90.

Proteins	$P_{weight}(w_j)$	CV	$C_{weight}(w_j^*)$	$\hat{\beta}_j$
FGF 5	1.00	0.39	1.00	0.0101
GASP	0.95	0.41	0.99	0.0047
CD62L	0.84	0.32	0.97	-0.0089
ICAM	0.85	0.41	0.97	0.0314
PDGF	0.82	0.50	0.95	-0.0062
EOTAXIN	0.83	0.72	0.94	-0.0045

TABLE 7.4: Number of selected features and AIC values.

Re-sampling Method	Number of features	AIC
C_{weight}	12	120.22
P_{weight}	3	165.45

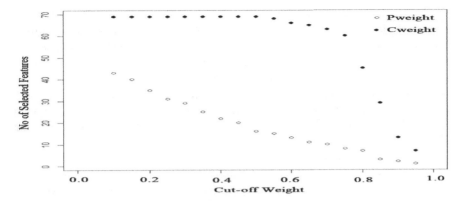

FIGURE 7.1: Number of features selected under P_{weight} and C_{weight} for different cut-off weights.

C_{weight} calculates the estimated hazard rate of relapse, and lower AIC represents a better model than the one which uses P_{weight}. Also, figure 7.1 shows that C_{weight} can select more features than the P_{weight} at the same cut-off weight. Therefore, this article illustrates a new feature selection method from a large dataset to run a predictive model.

The procedure on re-sampling technique shows a modified feature selection approach from high-dimensional data to handle any predictive model. The given example illustrates the usefulness of the modified weight function to develop a good model by choosing only 12 proteins from a set of 507. In this method, the training and validation datasets are randomly allocated once. As a result, we can overcome the biasedness owing to this random allocation only a single time.

7.4 Working with Training Datasets

The training datasets play an essential role in any feature selection method. It is investigated that embedded methods enormously rely on the taste of the training set and choose the features automatically based on some regularization functions [122, 123]. Moreover, the re-sampling algorithm does not like any elements from the training set by default but makes an ordering of the features by their relevance in the model [119]. The parts are then adjusted and selected in the ultimate model by selecting a threshold for the modified weight w_j^*. Therefore, we can observe that the C_{weight} function can eliminate less informative proteins more consistently than the previously used P_{weight} function as stated in [119].

A current study on high-dimensional data analysis uses the Bayesian concept on a group of generalized ridge regularization. This approach modifies the shrinkage using prior information of the regression coefficients [124]. On the contrary, the derived re-sampling algorithm does not choose any variables according to their estimate of the parameters. Before fitting the model, we first identify the variables based on the modified weight. And then, we only take the relevant components to the ultimate model. Hence, one can perform manual incorporation of any additional variables in the final model by relaxing the cut-off value.

In clinical research, sometimes the investigator prefers to control what is needed to include in the ultimate model than what is chosen by the method. The re-sampling method and the modified weight function give the investigator more control for developing the model. However, the choice of training set influences the final group of variables in the existing embedded methods [125]. Those methods may choose a different list of variables if we use an additional training or testing dataset or change the random sample proportions. The re-sampling algorithm and the modified weight function can now resolve this issue.

In the application of microbiome data, the goal is to identify the actual underlying biological signal. Reproducibility criteria for evaluating feature selection methods such as Nogueira's Stability measurement can help to determine the optimal feature selection method in this context [126, 127]. In contrast, the proposed method removes the matter of lacking reproducibility in the random selection of training and testing groups by considering the consistency of the variables and sorting criteria based on the steps of the algorithm. Recent advancement in genomics is bio-marker modeling that uses exquisite computational modeling methods like joint longitudinal modeling, Bayesian state-space modeling, time to event modeling, etc [128, 129, 130]. One can treat the re-sampling form and the modified weight function to identify the candidate features for these models. The prediction accuracy of these models can be improved by the re-sampling method. Furthermore, we require research on these advanced predictive models to measure the influence of different feature selection algorithms.

The above results display that the re-sampling method can produce a compelling selection of variables based on the new weight function. Still, we need a high programming computer and an amount of significant time due to re-sampling. Any further study on this method can evaluate its effect to solve other biases existing in microarray feature selection methods, such as multiple testing, imbalance class, normalization, etc [125, 131].

There are some limitations of the proposed approach. It is known that the coefficient of variation can be calculated only for the variables; it is either discrete or continuous, not for the attributes. In this consequence, the proposed re-sampling strategy does not take care of the categorical type variables. Since we couldn't calculate the CV for any attributes, the construction of the new weight function is improper for these types of factors. The choice of cut-off weight is another limitation of the proposed algorithm. Specification of the threshold is always in the hand of the researcher, and its proper choice can lead toward building a good prediction model and decreasing the AIC value.

There are different aspects in the analysis of cancer relapse. The essential task in the clinical setting is to supervise relapse-free survival and related outcomes. The time to relapse is the outcome variable in those similar situations. It will be more complicated to analyze the problems under the presence of other competing events such as death [132]. We can also use our algorithm for analyzing the time to event outcome by choosing suitable models like the logistic model, non-proportional hazards model. Assessment of the re-sampling algorithm together with the modified weight function under the different model building can be a topic for future research. Illustration of the modified method on weight function and the re-sampling algorithm provides us a new methodological way to select consistent features from a large dataset. This article also compares the new process and the existing one, which helps a researcher perform the predictive model. In addition, the example exhibits the usefulness of the proposed approach for developing an accurate model with a low AIC value by selecting only 12 out of 507 proteins. This new approach will help us overcome the challenges of biasedness in determining training and testing sets. However, we have required adequate time and good computer systems to run the algorithm properly for a high-dimensional dataset. The obtained results from the new approach still interpret the solutions more practical and provide enough space for the researcher to interact and choose features according to their importance.

7.5   Variable Selection in High-Dimensional Data

If the sample size is n of a study for vector $Y_i \in R^p, i = 1, 2, \ldots, n$, from k clusters with $p > n$ variables. The challenge is to group observation with k with clusters. The clusters are formed with homogenous values than the different ones. There are different statistical models proposed by high dimensional clustering problem [133, 134, 135, 136, 137, 138, 139, 140, 141, 142]. The primary way is to add the regularization method to create the sparsity [143] added an L_1 penalty on the cluster mean of each feature [137] by the pairwise group-fusion with the inclusion of penalty function to reduce the difference between the groups [140]. We create the sparse k-means and sparse hierarchical clustering with sparse weighted loss of each variable. Different numerical results of these methods are useful with the theoretical justification of these methods. Different clusters on latent space via matric factorization with random projection [135]. Another way to address the high dimensionality is through feature selection methodology [133, 134, 135, 136, 137, 138, 139, 140, 141, 142]. Several high-dimensional feature selection is suited to the supervised learning method.

Different unsupervised learning methods by Principal Component Analysis (PCA), by deriving the influential feature selection of Kolmogorov–Smirnov (KS) scores with sparse Gaussian mixture model.

However, cluster formation after having a selected variable is difficult to work due to convergence issues. The computational cost of forming the KS score is a challenging task. Recent computational efficiency development has created an optimal method to solve high-dimensional clustering. This method is motivated by recent progress in RNA sequencing by pseudo labeling techniques [144] with informative features. Initially, spectral clustering is formed. After that, select informative features by univariate regression on estimated labels. Finally, perform the spectral clustering with Lloyd's iteration method. It is possible to have a proposed algorithm that can successfully identify all informative features. The features are selected by the probabilistic method with a sparse Gaussian mixture model to perform the optimal misclustering rate [145]. This three-stage algorithm works through signal-to-noise ratio (SNR) and sample size calculation. There is a faster convergence rate of spectral clustering. The error rate of spectral clustering formed by $O(\sqrt{p/n})$ to $O((p/n))^{1/4}$ when $p > n$ [146]. The method can provide a way to work efficiently by characterizing the sub-populations from different clusters. Now identify the informative genes and gain a natural look into the high-dimensional data by RNA seq data.

7.6   Zellner's g-prior in Variable Selection

Zellner's g-prior is a Bayesian prior distribution for linear regression coefficients that is often used in variable selection problems. The g-prior is a generalization of the standard normal prior, and it has a parameter g that controls the amount of shrinkage applied to the coefficients. A larger value of g corresponds to more shrinkage and a smaller value corresponds to less shrinkage.

In variable selection, the g-prior can be used to balance the trade-off between model complexity and goodness-of-fit. A smaller value of g will result in less shrinkage, which will allow more variables to be included in the model, while a larger value of g will result in more shrinkage and fewer variables will be included in the model.

In R, the package R2jags can be used to implement Zellner's g-prior for variable selection in linear regression models. The package provides a function for fitting Bayesian linear regression models using JAGS, which is a program for Bayesian inference using Markov Chain Monte Carlo (MCMC) methods. To use the g-prior, you will need to specify the prior distributions for the parameters of the model and the transition probabilities between models.

It should be noted that Zellner's g-prior is not the only way to implement Bayesian variable selection, other alternatives include the spike and slab priors or the Bayesian Lasso.

7.7 Reversible Jump Markov Chain Monte Carlo

Reversible Jump Markov Chain Monte Carlo (RJMCMC) is a statistical method used for model selection and parameter estimation in Bayesian inference. It allows for the exploration of multiple models with different numbers of parameters, while maintaining the correct posterior distribution.

In R, the package rjags can be used to implement RJMCMC. It is an interface to JAGS (Just Another Gibbs Sampler), which is a program for Bayesian inference using MCMC methods.

Additionally, there's the package rjmcmc that provides a generic implementation of Reversible Jump MCMC for any model that can be specified in R.

When using RJMCMC, you will need to specify the prior distributions for the parameters of each model and the transition probabilities between models. You will also need to write the code for the likelihood function that calculates the probability of the data given the parameters of the model.

RJMCMC can be useful in situations where there is uncertainty about the number or structure of the parameters in a model, or when there is a need to compare multiple models. However, it can be computationally intensive and may require a significant amount of computational resources [147, 148, 149, 150, 151, 152].

7.8 Stochastic Variable Selection

Stochastic variable selection is a method for selecting a subset of variables from a large set of potential predictors in a statistical model. The selection is done in a random or stochastic manner, as opposed to deterministic methods such as stepwise selection.

There are several stochastic variable selection methods that can be used in R, including:

Random subspace method: This method selects a random subset of variables at each iteration and builds a model using those variables. The process is repeated multiple times and the variables that are selected most frequently are considered to be important predictors.

Random forest: The random forest algorithm is an ensemble learning method that builds multiple decision trees and selects a random subset of variables at each split. The variables that are selected most frequently across the trees are considered to be important predictors.

Bootstrap aggregating (bagging): This method builds multiple models using bootstrapped samples of the data and a random subset of variables. The variables that are selected most frequently across the models are considered to be important predictors.

Stochastic gradient boosting: this method is an optimization algorithm that uses a gradient descent optimization algorithm to minimize the loss function. It starts by building a simple base model and at each iteration it fits a new model to the negative gradient of the loss function.

The choice of method will depend on the specific goals and characteristics of the data. The R packages randomForest, caret, xgboost and gbm have functions for performing these techniques. [153].

7.9   MCMC Variable Selection

MCMC (Markov Chain Monte Carlo) variable selection is a method that can be used in Bayesian frameworks to perform variable selection in high-dimensional problems. The method uses MCMC techniques to generate a set of samples from the posterior distribution of the parameters, and it can be used to determine which variables are relevant and which can be excluded from the model.

The basic idea behind MCMC variable selection is to construct a model with all possible predictors and then use MCMC to explore the parameter space and identify the subset of predictors that are most important for explaining the outcome. The process is repeated multiple times to generate a set of samples from the posterior distribution of the parameters, which can then be used to estimate the marginal likelihood for each predictor.

In R, there are several packages that can be used to perform MCMC variable selection, such as "BMS", "MCMCglmm" and "rjags".

The "BMS" package provides a function "bms()" which can be used to perform Bayesian Model Averaging (BMA) using MCMC techniques. BMA is a type of MCMC variable selection that allows the model to select relevant predictors among the high dimensional data.

The "MCMCglmm" package provides a function "MCMCglmm()" which can be used to fit mixed-effects models using MCMC techniques. The package also includes variable selection methods based on MCMC such as the reversible jump MCMC (RJ-MCMC) algorithm.

The "rjags" package provides a function "jags()" which can be used to run JAGS (Just Another Gibbs Sampler) for Bayesian models. JAGS is a program for running Bayesian inference using MCMC methods. The "rjags" package allows the user to interface with JAGS from R, and it can be used to perform variable selection in high-dimensional problems by specifying the appropriate model and priors in the JAGS code and running the MCMC algorithm.

It's important to note that MCMC variable selection is a complex method and it's recommended to consult with experts in Bayesian statistics and variable selection to ensure that the analysis is correctly performed. Additionally, it's important to pay attention to the quality of the MCMC chains, such as checking for convergence, mixing and autocorrelation, and to adjust the MCMC algorithm accordingly if necessary.

7.10   Gibbs Variable Selection

Gibbs variable selection is a method for variable selection in high-dimensional data. It is a type of Bayesian variable selection method that uses MCMC sampling to explore the space of all possible models.

The basic idea behind Gibbs variable selection is to iteratively sample from the posterior distribution of the model parameters and the model structure (i.e., which variables are included in the model). At each iteration, the algorithm updates the model parameters given the current model structure and then updates the model structure given the current parameter estimates. This process is repeated many times to generate a sample of models from the posterior distribution.

The package BVS in R provides an implementation of the Gibbs variable selection method. It provides the BVS() function which can be used to fit a Bayesian variable selection model. The function takes a formula as input and the model should include the response variable as the left-hand side of the formula and the predictor variables as the right-hand side of the formula.

For example, if the data is in a dataframe called 'data' and the response variable is called 'y', the predictor variables are 'x1', 'x2', 'x3' and the number of iterations is 'niter', the Gibbs variable selection model can be fit using the following code:

```
library(BVS)
model=BVS(y ~ x1 + x2 + x3, data = data, niter = niter)
```

Once the model is fit, the summary() function can be used to obtain the estimates of the parameters, standard errors, and p-values, and the predict() function can be used to predict the response variable for new data.

It is important to note that, the choice of the model structure and the number of iterations will depend on the type of data, the pattern of missing data, and the research question. It is also important to check the assumptions of the model and the quality of the model using graphical methods and statistical tests before interpreting the results. [154, 153, 155].

7.11   High-Dimensional Data Analysis Using R

There are several techniques for analyzing high-dimensional data in R, including:

Principal Component Analysis (PCA): This is a technique used for dimensionality reduction and visualization of high-dimensional data. It finds the directions of maximum variance in the data and projects the data onto a lower dimensional space.

Multidimensional Scaling (MDS): This is a technique used for visualizing high-dimensional data in a lower-dimensional space, similar to PCA. However, unlike PCA, MDS tries to preserve the pairwise distances between data points.

Factor Analysis: This is a statistical technique used to identify underlying patterns in a high-dimensional dataset. It finds a smaller number of latent variables that can explain the variation in the data.

Clustering: Clustering is a technique used for grouping similar observations together in a high-dimensional dataset. Popular clustering algorithms for high-dimensional data include density-based clustering and subspace clustering.

t-SNE: t-SNE (t-Distributed Stochastic Neighbor Embedding) is a dimensionality reduction technique that is particularly well suited for visualizing high-dimensional data in a 2D or 3D space.

The choice of technique will depend on the specific goals and characteristics of the data. The R packages principal, MASS, psych, cluster, factoextra, and Rtsne have functions for performing these techniques.

```
Creates a network plot of high dimensional variables and lists those
variables

data(hnscc)
hdClust(7,105,0.05,2,ID="id",OS="os",
Death="death",PFS="pfs",Prog="prog",hnscc)
```

```
LASSO for high dimensional data

data(hnscc)
hidimLasso(7,105,OS="os",Death="death",hnscc)
```

Survival analysis on high dimensional data

```
data(hnscc)
hidimSurv(7,105,0.05,ID="id",OS="os",
Death="death",PFS="pfs",Prog="prog",hnscc)
```

Survival analysis on high dimensional data for significant variables on OS and survival event

Variables	HR[LCL,UCL]	P-value
LRP2	1.09[1.00,1.19]	0.036
C2orf72	1.07[1.00,1.14]	0.049
GAL3ST1	1.13[1.03,1.24]	0.007
ZMAT1	0.89[0.84,0.96]	0.001
AFF3	0.88[0.82,0.95]	0.001
METTL7B	1.09[1.02,1.16]	0.006
PCDHA10	1.09[1.01,1.17]	0.023
C1orf88	1.10[1.02,1.18]	0.007
KIF1A	1.07[1.03,1.12]	0.000
CTSE	0.91[0.85,0.98]	0.021
CDHR5	0.84[0.72,0.97]	0.024
B3GAT1	0.90[0.82,0.98]	0.023
RBP4	1.07[1.02,1.13]	0.002
PGC	1.14[1.02,1.28]	0.021
MLXIPL	1.06[1.01,1.12]	0.014

Survival analysis on high dimensional data for significant variables on PFS and progression event

Variables	HR[LCL,UCL]	pvalue
ZMAT1	0.88[0.80,0.96]	0.006
AFF3	0.88[0.79,0.97]	0.012
METTL7B	1.09[1.00,1.18]	0.039
CHDH	0.92[0.85,0.99]	0.039
B3GAT1	0.88[0.78,0.99]	0.043
RBP4	1.08[1.01,1.15]	0.015
PGC	1.18[1.02,1.36]	0.020

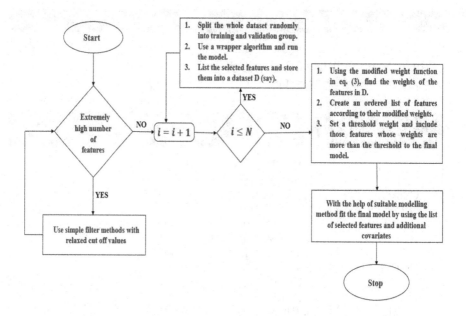

FIGURE 7.2: High-dimensional data varaible selection.

Survival analysis on high dimensional data for the DEGs/Variables found common between significant

"Estimates data for the DEGs/Variables found common between significant
DEGs from data having death due to progression and data showing death without progression"

Variables	HR.x[LCL.x,UCL.x]	pval-x	HR.y[LCL.y,UCL.y]	pval-y
AFF3	0.88[0.82,0.95]	0.001	0.88[0.79,0.97]	0.012
B3GAT1	0.90[0.82,0.98]	0.023	0.88[0.78,0.99]	0.043
METTL7B	1.09[1.02,1.16]	0.006	1.09[1.00,1.18]	0.039
PGC	1.14[1.02,1.28]	0.021	1.18[1.02,1.36]	0.020
RBP4	1.07[1.02,1.13]	0.002	1.08[1.01,1.15]	0.015
ZMAT1	0.89[0.84,0.96]	0.001	0.88[0.80,0.96]	0.006

Uses bonferroni correction factor in survival analysis

```
data(hnscc)
hidimSurvbon(7,105,0.05,ID="id",OS="os",
Death="death",PFS="pfs",Prog="prog",hnscc)
```

Two step fileration using bongerroni correction and LASSO

```
data(hnscc)
hidimSurvbonlas(6,104,0.05,ID="id",OS="os",
Death="death",PFS="pfs",Prog="prog",hnscc)
```

Two step verification without bonferroni correction

```
data(hnscc)
hidimSurvlas(7,105,0.05,ID="id",OS="os",
Death="death",PFS="pfs",Prog="prog",hnscc)
```

Results on bonferroni correction

"AFF3" "B3GAT1" "METTL7B" "PGC" "RBP4" "ZMAT1"

Creating batches of variables on high dimensional data

```
data(hnscc)
hidimsvc(7,105,5,0.05,ID="id",OS="os",
Death="death",PFS="pfs",Prog="prog",hnscc)
```

Result on batch process outcomes

"Estimate values of significant variables/DEGs on considering Death with Progression"

Variables	HR[LCL,UCL]	pvalue
LRP2	1.09[1.00,1.19]	0.036
C2orf72	1.07[1.00,1.14]	0.049
GAL3ST1	1.13[1.03,1.24]	0.007
ZMAT1	0.89[0.84,0.96]	0.001
AFF3	0.88[0.82,0.95]	0.001
METTL7B	1.09[1.02,1.16]	0.006
PCDHA10	1.09[1.01,1.17]	0.023
C1orf88	1.10[1.02,1.18]	0.007
KIF1A	1.07[1.03,1.12]	0.000
CTSE	0.91[0.85,0.98]	0.021
CDHR5	0.84[0.72,0.97]	0.024
B3GAT1	0.90[0.82,0.98]	0.023

Chapter 8

Bayesian Survival Mediation Data Analysis

8.1 Introduction

Conventional mediation analysis methods poorly serve the dual objectives and detect the difference between complete and partial mediation by the matrix. The classical approach is established by the mediator with X Matrix [1]. It is commonly known as the causal steps(CS) method. It is established by the relationship between X, M, and Y. Sometimes, CS applies by linear regression model to perform with four conditions: (1) X has a marginal effect on Y. (2) X has an effect on M conditions; (3) M is at least a partial mediator of the impact of X on Y; and (4) M is a complete mediator of the effect of X on Y. The CS method is used to work with the matrix of independent variables by the likelihood ratio test for the grouped predictors. The CS is beneficial in conventional data analysis but challenging to work with high-dimensional data analysis. It can work with multiple testing due to option (4) above. It is challenging to summarize the evidence for partial or complete mediation for candidate M.

8.2 Bivariate Survival Model

A bivariate survival model is a statistical model that is used to analyze the relationship between two survival outcomes, such as time-to-death or time-to-failure, in a single population. The goal of a bivariate survival model is to estimate the joint distribution of both survival outcomes, and to identify factors that are associated with the risk of both events occurring.

There are several different types of bivariate survival models, such as:

Copula models: These models use a copula function to model the dependence between the two survival outcomes. Joint models: These models model the joint distribution of the two survival outcomes, and use a shared set of covariates to model the dependence between the outcomes. Multi-state models:

DOI: 10.1201/9781003298373-8

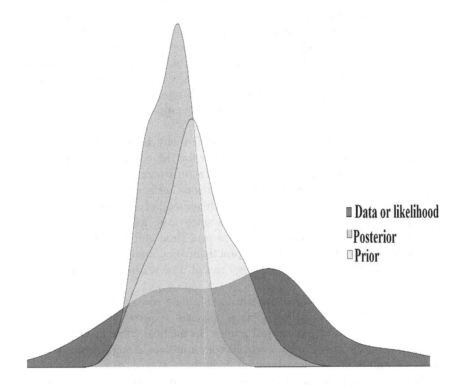

FIGURE 8.1: Bayesian Prior, Likelihood and Posterior.

These models use a Markov process to model the transitions between different health states, and use a shared set of covariates to model the dependence between the outcomes. Bivariate survival models are particularly useful when there is a clear interdependence between the two survival outcomes, and when it is important to estimate the joint distribution of the outcomes. They are widely used in fields such as medicine, epidemiology, and engineering to analyze time-to-event data, and to identify factors that are associated with the risk of both events occurring.

8.3   Bayesian Proportional Hazard Model

Coherent learning from experimental data is the foundation of Bayesian theory. It is an inductive inferential strategy that considers the observed result and expands it to account for population-wide uncertainty. In contrast to the Bayesian approach, the classical approach uses a deductive process to draw conclusions about the population parameters. The Bayesian is willing to gather data from various sources and make assumptions about the population. The Bayesian methodology differs from the traditional method in drawing conclusions and combining evidence. Combining the evidence from different sources to produce the output is also beneficial. The Bayesian from the observed outcome considers the inductive approach, enlarging the assertion for the general population.

Clinical trials for cancer can be carried out both retrospectively and prospectively. The stopping rule is crucial for the continuation of the study in the event of a potential issue. The decision to continue the study can be thought about based on the trial's information that is now accessible. Through Bayesian analysis, the estimations of the likelihood of success can be obtained. The proactive strategy of ending the practice can help patients avoid needless exposure to hazardous medications and time and money. The Bayesian approach is easier to use and more practical when analyzing the current posterior distribution value. The strategy can also be taken into account for data monitoring tools.

Moment-to-moment or in order to manipulate the outcome variable where an interesting occurrence takes place, survival is presented. This technique is used elsewhere and has several names. In medical, they are referred to as survival analysis; in sociology, event history analysis; in engineering, reliability analysis; and in economics, duration analysis.

However, an observational unit is always used to measure the survival analysis. Due to left-wing filtering, the event time is not recorded. Right and interval censoring are other forms of censoring. The combination of time and censoring separates the survival analysis from other forms of analysis. In general, two main sorts of research—firstly the instantaneous rate of the event and secondly as the function of time—can be used to complete the survival analysis. A proportional and non-proportional hazard regression model were among the models. Accelerated failure time (AFT) models can occasionally be used with models. The recurring event can be handled by a competing risk model. Cure rate modeling, cluster survival modeling, and longitudinal and Survival modeling are further forms of extension of survival models. However, a variety of software programs have been created to use with survival analysis. However, the majority of those use conventional methods. There hasn't been much advancement in Bayesian survival analysis study. Bayesian survival analysis is difficult to use with traditional data analysis software. In a same vein, Bayesian Survival-specific software is not generally used for other sorts of data analysis. Therefore, there is relatively little approval for their widespread use. There are some recent developments to solve the challenge by adopting Bayesian in Cox PH with SurviMChd package in R. It is available to perform Cox PH through the Bayesian approach. The function available in the SurviMChd package named "survMC" is suitable to work with the Cox Proportional Hazard (PH) function in R.

8.4 Bayesian Smoothing Methods

Bayesian smoothing methods are a class of statistical techniques that use Bayesian principles to estimate smooth functions or curves from data. These methods are used to estimate unknown quantities from observed data, by expressing prior beliefs about the quantity of interest as a probability distribution.

There are several different types of Bayesian smoothing methods, such as:

Bayesian splines: These methods use a set of basic functions and a smoothing prior to model the function of interest. Gaussian Processes (GPs): These methods use a Gaussian prior on the function of interest, and a covariance function to model the smoothness of the function. Markov Chain Monte Carlo (MCMC) methods: These methods use a Markov chain to sample from the posterior distribution of the function of interest. Bayesian smoothing methods are particularly useful when the data is noisy or when the functional form of the relationship is not known. They are widely used in fields such as signal processing, image processing, and machine learning to estimate smooth functions, and in fields such as bioinformatics and pharmacokinetics for time-series data analysis.

8.5 Nonparametric Bayesian Model

Nonparametric Bayesian models are a class of statistical models that use Bayesian principles to estimate complex and flexible functions or distributions from data without making strong assumptions about the functional form of the relationship. These models are also called "semi-parametric" or "nonparametric" models.

There are several different types of nonparametric Bayesian models, such as:

Dirichlet Process Models (DPMs): These models use a Dirichlet process as a prior distribution to model the distribution of the data without specifying the number of clusters or components. Gaussian Processes (GPs): These models use a Gaussian prior on the function of interest, and a covariance function to model the smoothness of the function without specifying the functional form of the relationship. Chinese Restaurant Process (CRP) Models: These models use a CRP as a prior distribution to model the number of clusters or components in the data without specifying the number of clusters or components. Nonparametric Bayesian models are particularly useful when the functional form of the relationship is not known or when the data is complex and high-dimensional. They are widely used in fields such as machine learning, computer vision, natural language processing, and bioinformatics. In R, the package DPpackage provides functions for fitting nonparametric Bayesian models using the Dirichlet Process (DP) prior. The DP prior is a flexible prior distribution that can be used for density estimation and clustering.

Another package is bnlearn which can be used to fit Bayesian networks, a class of graphical models that can be used for both classification and regression tasks. It provides an extensive collection of algorithms for learning the structure of Bayesian networks, as well as functions for fitting models.

The package hdp is another alternative, it can be used to fit Hierarchical Dirichlet Process (HDP) models, a nonparametric Bayesian method for clustering and topic modeling.

Finally, there is the package stancode which provides a flexible framework for fitting a variety of nonparametric Bayesian models using the Stan probabilistic programming language.

It should be noted that nonparametric Bayesian models are computationally intensive and may require a significant amount of computational resources. Furthermore, choosing the right model depends on the specific goals and characteristics of the data.

8.6　The Dirichlet Process Prior

The Dirichlet Process (DP) prior is a nonparametric Bayesian prior distribution that can be used for a wide range of applications, including density estimation, clustering, and Bayesian nonparametric regression.

The DP prior is a probability distribution over distributions. It is defined by a base distribution G_0 and a concentration parameter alpha. The DP prior generates a sample distribution G from an infinite mixture of distributions, each of which is a copy of G_0 with a different set of parameters. The sample distribution G can be thought of as a probability distribution over an infinite set of discrete clusters.

The DP prior has several useful properties, including:

It is a conjugate prior for a wide range of models, including the Chinese restaurant process and the Polya urn scheme. It allows for the incorporation of prior knowledge about the underlying distribution. It can be used to automatically infer the number of clusters in a dataset, without the need to specify the number of clusters beforehand. The DP prior is typically used as the prior distribution in a Bayesian model, in conjunction with likelihood function that describes the data. Inference is typically done using MCMC methods, such as Gibbs sampling, or variational inference.

In R, the package DPpackage provides functions for fitting nonparametric Bayesian models using the Dirichlet Process (DP) prior. Additionally, the package hdp can be used to fit Hierarchical Dirichlet Process (HDP) models, which are a generalization of the DP model that allows for modeling hierarchically structured data.

8.7   Dirichlet Process (DP) Prior Using R

The Dirichlet Process (DP) prior can be used in R to perform nonparametric Bayesian modeling, such as density estimation, clustering, and Bayesian nonparametric regression.

The package DPpackage provides functions for fitting nonparametric Bayesian models using the DP prior. It includes functions for:

Density estimation using the DP mixture model clustering using the Chinese restaurant process; Bayesian nonparametric regression using the DP Gaussian process; To use the DP prior in a model, you will need to specify the base distribution G_0 and the concentration parameter alpha. The package provides several options for the base distribution, including the Gaussian, Poisson, and Exponential distributions. The concentration parameter controls the trade-off between the prior and the data. A larger value of alpha will result in more weight being placed on the prior, while a smaller value will result in more weight being placed on the data.

Inference is typically done using MCMC methods, such as Gibbs sampling. The package provides functions for performing MCMC inference, as well as functions for summarizing and visualizing the results.

It should be noted that using the DP prior can be computationally intensive and may require a significant amount of computational resources. Additionally, the choice of base distribution and concentration parameter will depend on the specific goals and characteristics of the data.

8.8   Bayesian Mediation Analysis

Bayesian Mediation Analysis (BMA) is a statistical method that combines the principles of Bayesian statistics and mediation analysis to estimate the causal effect of an independent variable on a dependent variable through an intermediate variable, known as a mediator variable. The goal of BMA is to identify and understand the underlying mechanisms or processes that link an independent variable to a dependent variable, and to make inferences about these relationships based on observed data.

BMA uses Bayesian methods to estimate the unknown parameters of the model, such as the strength and direction of the causal relationships between the variables, and to quantify the uncertainty in these estimates through the use of prior distributions and posterior distributions. BMA also allows for the incorporation of prior knowledge and information into the analysis.

BMA is increasingly used in fields such as psychology, marketing, and economics to understand how certain factors influence outcomes. It can be used to determine the extent to which a mediator variable explains the relationship between an independent and dependent variable, and to make predictions about the effect of interventions on the outcome variable of interest.

8.9 Bayesian Survival Using R

Bayesian survival analysis using R is a statistical method that uses the R programming language to perform Bayesian analysis of time-to-event data. R is a widely used open-source programming language and environment for statistical computing and graphics. It has a rich ecosystem of packages and libraries that can be used to perform Bayesian survival analysis.

There are several R packages that can be used to perform Bayesian survival analysis, such as: "rstanarm": This package provides a convenient interface to perform Bayesian survival analysis using the Stan software, which is a powerful engine for Bayesian inference. "brms": This package provides a high-level interface for fitting Bayesian generalized linear mixed models using the Stan software. "JAGS": This package provides an interface to the JAGS software, which is another engine for Bayesian inference. These packages allow for the estimation of survival models with various distributions of the time-to-event variable, such as Weibull, exponential, or lognormal, and different forms of regression models, such as parametric, semiparametric, or nonparametric. They also provide tools for model comparison, model checking, and visualization of the results. Bayesian survival analysis using R provides a flexible and powerful framework for analyzing time-to-event data, and for making inferences about the underlying risk factors and their effect on the outcome.

High dimensional survival analysis with interval censored data by MCMC

```
data(hnscc)
survintMC(m=7,n=11,Leftcensor="leftcensoring",
OS="os",Death="death",iter=6,data=hnscc)
```

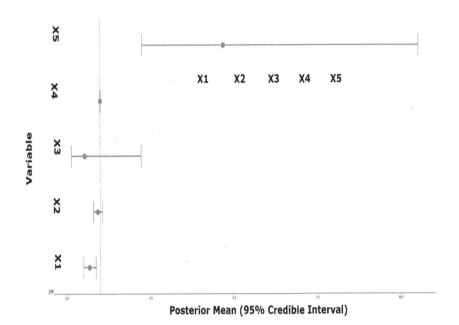

FIGURE 8.2: Posterior Mean estimates though survMC function.

Results on high dimensional survival analysis with interval censored data by MCMC		
Variables	HR	[LCL,UCL]
GJB1	1.00	[1.00,1.00]
PPP1R9A	0.90	[0.89,0.90]
HPN	0.95	[0.95,0.95]
SLC4A4	0.94	[0.93,0.95]
HNF1B	1.00	[0.99,1.00]

Survival analysis using Cox Proportional Hazards with MCMC

```
data(mcsurv)
survMC(m=4,n=8,Time="OS",Event="Death",
chains=2,adapt=100,iter=1000,data=mcsurv)
```

Result on Cox Proportional Hazards with MCMC

Variables	Posterior Means(SD)	[2.5%,97.5%]	DIC
x1	0.67(1.15)	[0.50,0.87]	219.06
x3	0.93(1.07)	[0.81,1.07]	227.15
x2	0.54(2.00)	[0.14,2.22]	227.36
x4	1.00(1.00)	[0.99,1.02]	226.38
x5	4.68(1.50)	[2.24,10.54]	212.86

Survival analysis on multiple variables with MCMC

```
data(mcsurv)
survMCmulti(var1="x1",var2=NULL,var3="x3",var4="x2",
var5="x4",Time="OS",Event="Death",chains=2,adapt=100,
iter=1000,data=mcsurv)
```

Result on multiple variable test with MCMC

Variables	Posterior Means(SD)	[2.5%,97.5%]
x1	0.67(1.20)	[0.46,0.93]
x3	0.86(2.24)	[0.16,3.79]
x2	0.94(1.08)	[0.80,1.09]
x4	1.00(1.00)	[0.98,1.01]

Chapter 9

Bayesian Accelerated Failure Time Mediation Data Analysis

9.1 Introduction

Bayesian accelerated failure time (AFT) mediation analysis is a statistical method that combines the principles of Bayesian statistics, AFT models, and mediation analysis to estimate the causal effect of an independent variable on a dependent variable through an intermediate variable, known as a mediator variable. The goal of Bayesian AFT mediation analysis is to identify and understand the underlying mechanisms or processes that link an independent variable to a dependent variable, and to make inferences about these relationships based on observed data.

AFT models are used to estimate the effect of an independent variable on the time to an event, such as death or failure, rather than the event itself. Bayesian AFT mediation analysis uses Bayesian methods to estimate the unknown parameters of the AFT model, such as the strength and direction of the causal relationships between the variables, and to quantify the uncertainty in these estimates through the use of prior distributions and posterior distributions.

Bayesian AFT mediation analysis is particularly useful when the data is time-to-event data and when the goal of the analysis is to understand the causal relationship between an independent variable, a mediator variable, and a time-to-event dependent variable. It is widely used in fields such as medicine, epidemiology, and engineering to analyze time-to-event data, and to identify factors that are associated with the risk of an event occurring.

9.2 Accelerated Failure Time in High-Dimensional Survival Data

Let T be the 'failure time', and the outcome is x, the corresponding covariate vector. We can prepare the statistical inference by exploring the effect from x on the outcome variable T. Without censored observation, we can regress T by transforming it directly on the covariate x. This approach is attractive for practice and interpretation. If censored observation is present in the data, we can adopt the conventional approach to explore the covariate effect. Sometimes we exclusively use the proportional hazards (PH) model by the likelihood approach to draw inferences about the covariate effect. This approach is relatively flexible. It is appropriate to work with time-dependent covariates. Therefore the PH model defines the effect of the covariate x as it acts multiplicatively on the hazard function. It is not easy to interpret with the estimate of the regression parameters. The ordinary linear regression model can be handled with censored observations. It is useful as an alternative to the Cox model in the time-to-event data analysis. The presence of censored observation, the linear regression with $\log T$ as the response variable, has been explored extensively.

The AFT is a regression model. It is related to the log-event times as a linear function of the covariates. The AFT assumes that the covariates decelerate or accelerate the expected event time. Sometimes the cluster correlation can be modeled by the AFT model by random effects by the classical linear mixed models. Sometimes it is desirable to estimate the parametric distribution by normal or exponential distribution. The Bayesian method is flexible to work with censored observation with accurate event times as latent variables in the model to explore the unknown parameters. This work can also be explored with the nonparametric distribution [156, 157, 158, 159, 160].

Let T_i be the failure time for the ith patient, $i = 1, ..., n$. For T_i, we can only observe a bivariate vector (Y_i, Δ_i), where $Y_i = \min(T_i, c_i)$ and $\Delta_i = 1$ if $T_i = Y_i$ and 0 otherwise. The c_i s are censoring variables. Conditional on the covariates for the ith subject, c_i is assumed to be independent of the failure times $T_i, i = 1, ..., n$. Let x_i denote a $p \times 1$ vector of covariates for T_i. Suppose that the base 10 logarithm of T_i is linearly related to x_i, that is, there exists an unknown constant β such that

$$logT_i = \beta^{'} x_i + e_i, \qquad (9.1)$$

where $e_i, i = 1, ..., n$, are independent and identically distributed random variables whose common distribution function F is completely unspecified. Note that other strictly increasing functions might replace the logarithm. The intercept parameter is not included in the vector β. Therefore, the mean of the error term may not be 0. The presence of censoring means the intercept parameter cannot be estimated well.

9.3   Bayesian Accelerated Failure Time in High-Dimensional Survival Data

Bayesian AFT models in high-dimensional survival data are a class of statistical models that are used to analyze time-to-event data in the presence of a large number of covariates. AFT models estimate the relationship between a response variable (time-to-event) and one or more predictor variables, while accounting for the high-dimensionality of the data.

In Bayesian AFT models, the likelihood function is based on the AFT distribution, which is a generalization of the exponential distribution. The AFT distribution allows for modeling of both right-censored and interval-censored data, which is common in survival analysis.

To handle the high-dimensionality of the data, this method can use various dimensionality reduction techniques such as variable selection method, shrinkage methods, or regularization methods to select a subset of relevant covariates. After selecting the relevant covariates, the model can be estimated using Bayesian methods such as Markov Chain Monte Carlo (MCMC) techniques, which allow for the incorporation of prior knowledge and information into the analysis.

This method is particularly useful when the goal of the analysis is to understand the relationship between a time-to-event response variable and one or more predictor variables, while accounting for the high-dimensionality of the data. They are widely used in fields such as medicine, epidemiology, and engineering to analyze time-to-event data in high-dimensional settings. The AFT model is a linear regression model in which the response variable is the logarithm or a known monotone transformation of a failure time [161]. As a useful alternative to the Cox model [162], this model has an intuitive linear regression interpretation [163] for a lucid discussion. Semiparametric estimation in the AFT model with an unspecified error distribution has been studied extensively in the literature for right censored data. In particular, two methods have received special attention. One method is the Buckley-James estimator which adjusts censored observations using the KaplanMeier estimator. The other is the rank-based estimator which can be motivated from the score function of the partial likelihood [164, 165, 166, 167] among others. However, the AFT model has not been widely used in practice, mainly due to the difficulties in computing the semiparametric estimators of the aforementioned methods, even in situations when the number of covariates is relatively small [168]. For high-dimensional covariates it is even more difficult to apply these methods, or their regularized versions, especially when variable selection is needed along with estimation.

9.4 Bayesian Accelerated Failure Time Mediational Analysis

Bayesian AFT mediational analysis can be performed using the R programming language and several packages available in R such as "rstanarm", "brms", "jmv", and "mediation". These packages provide functions for fitting AFT models and conducting Bayesian mediational analysis.

To perform AFT mediational analysis in R, you would first need to install and load the necessary packages. Then, you would need to prepare your data by specifying the treatment or exposure variable, the outcome variable, and the mediator variable. Once your data is prepared, you can use the functions in the packages to fit the AFT model and estimate the parameters of interest.

You can also use the functions to conduct sensitivity analyses, such as assessing the robustness of the results to different prior distributions and model specifications.

It is also important to note that to perform a Bayesian analysis, you will also need to specify prior distributions for the model parameters, and use MCMC methods to obtain posterior distributions for the parameters.

It's always good to have a good understanding of the statistical theory behind the models and the Bayesian framework before attempting to implement this type of analysis. It is also recommended to consult specialized literature and have help from a statistician if possible.

9.5 Illustration Using R

Accelerated failure time in high dimensional survival data using R

```
library(afthd)
data(hdata)
pvaft(9,30,STime="os",Event="death",0.1,hdata)
```

Result on accelerated failure time in high dimensional survival data using R

	Estimate	SE	Z-value	P-value
GAL3ST1	-0.13	0.05	-2.40	0.01
LRP2	-0.09	0.04	-1.95	0.05
NME5	-0.08	0.04	-1.79	0.07

Accelerated failure time in high dimensional survival data Using R

```
data(hdata)
set.seed(1000)
rglaft(9,50,STime="os",Event="death",1,hdata)
```

Bayesian accelerated failure time mediation data analysis using R

	Estimate	SE	Z-value	P-value
ZMAT1	0.11	0.03	2.91	0.00
GAL3ST1	-0.13	0.05	-2.40	0.01
LRP2	-0.09	0.04	-1.95	0.05
NME5	-0.08	0.04	-1.79	0.07
ADH1B	-0.04	0.02	-1.76	0.07
XIST	0.02	0.01	-1.68	0.09
BEX1	-0.06	0.03	-1.62	0.10
C2orf72	-0.06	0.03	-1.55	0.11
FAM47E	-0.06	0.04	-1.43	0.15
GOLT1A	-0.04	0.03	-1.29	0.19

Bayesian accelerated failure time in high dimensional survival data using R

```
library(afthd)
data(hdata)
set.seed(1000)
rglwbysm(9,45,STime="os",Event="death",2,10,1,hdata)
```

Accelerated failure time in high dimensional survival data using R

	Mean(SD)	[X2.5,X97.5]	Rhat	n.eff
alpha	0.93(0.06)	[0.83, 0.00]	1.81	4
beta[1]	7.13(0.13)	[6.87,0.00]	1.14	10
beta[2]	-0.01(0.04)	[-0.04,0.00]	1.67	4
beta[3]	-0.01(0.06)	[-0.08,0.00]	2.48	3
beta[4]	-0.08(0.05)	[-0.16,-0.00]	1.08	10
beta[5]	-0.07(0.04)	[-0.12,-0.00]	0.95	10
beta[6]	-0.06(0.07)	[-0.19,0.41]	3.71	2
Deviance	4006.99(35.79)	[3947.17,4060]	1.13	10
sigma	1.07(0.07)	[0.99,119]	1.81	4
tau	0.88(0.11)	[0.69,1.01]	1.81	4

Bayesian accelerated failure time mediation data analysis using R

```
library(autohd)
hdaftma(m=8,n=80,survdur="os",event="death",sig=0.05,
ths=0.02,b=10,d=10,data=hnscc2)
```

Bayesian accelerated failure time mediation data analysis using R

Bayesian Accelerated Failure Time (AFT) mediational analysis is a statistical method used to examine the effects of mediators on the time to event outcome in clustered survival data. This method is based on the AFT model, which is a semiparametric Cox proportional hazards model. Using this method, it is possible to estimate the effect of a mediator on the time to event outcome, as well as to determine whether there is a causal connection between the mediator and the outcome. Additionally, the Bayesian AFT mediational analysis can also be used to evaluate the impact of other covariates on the time to event outcome. This method is particularly useful in studies where the data are clustered and survival data are available, as it allows for better estimation of the effects of the mediator on the outcome. 'Active variables and their beta and alpha means'

Active_variables	beta.m	alpha.m
AFF3	0.98	0.12
KIF1A	-0.52	1.00

Chapter 10

Bayesian Competing Risk Mediation Data Analysis

10.1 Introduction

Bayesian competing risk mediation data analysis is a statistical method used to estimate the causal effects of a treatment on an outcome while accounting for the presence of competing risks. Competing risks refer to the presence of other events or outcomes that can occur and affect the outcome of interest. In this case, the competing risks are the other causes of death or failure.

In Bayesian competing risk mediation data analysis, the model is typically specified using a competing risks survival model, such as the Fine and Gray model. The model is then extended to include mediator variables and treatment variables. The model is then fitted using Markov Chain Monte Carlo (MCMC) methods, such as Gibbs sampling, to estimate the causal effects of the treatment on the outcome while accounting for the competing risks and the mediator variables.

In R, the package competingrisks can be used for fitting and analyzing competing risks data using Bayesian methods. It provides functions for fitting the Fine and Gray model, including the extended version for mediation analysis, and it also provides functions for summarizing and visualizing the results.

It is important to note that the analysis of competing risks data can be complex and requires a good understanding of the underlying assumptions and limitations of the models used. Therefore, it is important to carefully consider the assumptions of the model and to check the model fit and assumptions before interpreting the results.

Competing risks are common in time-to-event data analysis. There is a K number of causes for the events that may occur. The appearance of any of the risks might cause failure or death and precludes the occurrence of the different competing risks. It is allowed to work with only one competing risk data that allows observing only the failure time and a cause of failure for each unit of the study. But in reality, with medical conditions, there are two situations. One is the different causes of death. Secondly, it is common in cancer research, has as one competing risk recurrence of cancer and competing risk as death remission.

It is common to have recurrence, death, and work with competing risks in cancer and genomic research. The work with the interplay between the competing chances of recurrence and death raised the whole idea about treatment efficacy. The probability calculated with treatment cures and disease recurrence provides evidence of the treatment's toxicity. Sometimes, it requires maintaining the trade between the two competing risks with less toxic treatments. For illustration, lower doses of chemotherapy have a lower death risk but a high probability of recurrence.

In medical research, the objective is to explore the difference in outcomes due to treatment modalities (like chemotherapy, doses of chemotherapy) and adjust for other risk factors for disease outcomes. Different studies are presented to explore the risk factors for outcomes for a specific disease. There are different approaches to working with time-to-event data. Mostly these are related to Cox regression models for events like recurrence, death, and a combined event of therapeutic effects or disease-free survival (DFS). It is required to test the validity of the Cox model through time-dependent covariate modeling or performing the diagnostic plots.

It is essential to know that there are some challenges in predicting the effect of a factor on its treatment failure, recurrence, and death in remission. Sometimes the regression estimates resulting from relapse and treatment mortality differ in signs. It is difficult to predict the effect of the covariate on treatment failure without looking at the baseline cause-specific hazard rates for recurrence and treatment-related mortality. It commonly complicates the analysis. Indeed, the existing Cox model is not sufficient to be defined as the best model in this context. It might generate some misleading outcomes.

10.2   Competing Risk in High-Dimensional Survival Data

In high-dimensional competing risk data analysis, the number of covariates is much larger than the number of events. This presents a number of challenges, such as curse of dimensionality and overfitting. Bayesian methods can be used to address these challenges by incorporating prior information and performing variable selection.

One common approach to variable selection in high-dimensional competing risk data is to use shrinkage priors, such as the horseshoe or the Bayesian Lasso, which encourage small regression coefficients for unimportant variables. Another approach is to use variable selection methods that are specifically designed for competing risks data, such as the variable selection method for the Fine and Gray model.

In R, the package competingrisks provides functions for fitting the Fine and Gray model with shrinkage priors such as horseshoe and Bayesian Lasso. The package bayesSurv also provides a Bayesian variable selection method for Fine and Gray model which is specifically designed for competing risks data.

It is important to note that in high-dimensional competing risk data, it is crucial to use appropriate methods for variable selection and model assessment, such as cross-validation, stability selection, or other methods. Additionally, when using high-dimensional data, it is important to pay attention to the assumptions of the model and to check the model fit and assumptions before interpreting the results. Different variable selection techniques are evolved for high-dimensional small sample size time to event data [169, 170, 171, 172]. Microarray data raised many variables with small, effective sample size and can be selected reliably. It takes only on non-zero gene variables that are relevant for disease progression [173]. High-dimensional variable selection techniques like penalized methods such as least absolute shrinkage and selection operator (LASSO) [174, 175] and these are based on likelihood-based boosting methods [173]. High-dimensional data with highly-correlated variables raised the non-zero estimates [176]. Sometimes, the elastic net penalization covers this drawback in the time-to-event data [176]. The LASSO method works with automatic variable selection and continuous variable shrinkage. Sometimes the elastic net penalization is suitable for performing the grouped selection. It works with entire set of correlated genes [176, 86], while remaining computationally efficient [86]. This technique is suitable to work with single survival endpoint [173, 177, 178]. The likelihood-based approach for variable selection is adopted in the context of high-dimensional time-to-event competing risk data [173]. However, the performance of other methods like elastic net and Lasso for gene selection for competing risks data is not covered.

10.3   Bayesian Mediation Analysis in High-Dimensional Survival Data

Bayesian mediation analysis in high-dimensional survival data is a statistical method that combines the principles of Bayesian statistics, mediation analysis, and survival analysis to estimate the causal effect of an independent variable on a dependent variable through an intermediate variable, known as a mediator variable, in the presence of a large number of covariates. The goal of this method is to identify the underlying mechanisms or processes that link the independent variable to the dependent variable, and to make inferences about these relationships based on observed data while taking into account the large number of covariates.

To handle the high-dimensionality of the data, this method can use various dimensionality reduction techniques such as variable selection method, shrinkage methods, or regularization methods to select a subset of relevant covariates. After selecting the relevant covariates, the model can be estimated using Bayesian methods such as MCMC techniques, which allow for the incorporation of prior knowledge and information into the analysis.

This method is particularly useful when the goal of the analysis is to understand the causal relationship between an independent variable, a mediator variable, and a time-to-event dependent variable and the data has a high dimensionality. It is widely used in fields such as medicine, epidemiology, and engineering to analyze time-to-event data in high-dimensional settings.

10.4   Bayesian Competing Risk in High-Dimensional Survival Data

Bayesian competing risk analysis in high-dimensional survival data is a statistical method that combines the principles of Bayesian statistics and competing risk analysis to estimate the risk of different events occurring in the presence of a large number of covariates. Competing risk analysis is used when there are multiple types of events that can occur, each with its own unique set of risk factors.

In high-dimensional survival data, dimensionality reduction techniques can be used such as variable selection method, shrinkage methods, or regularization methods to select a subset of relevant covariates. After selecting the relevant covariates, the model can be estimated using Bayesian methods such as MCMC techniques, which allow for the incorporation of prior knowledge and information into the analysis.

This method is particularly useful when the goal of the analysis is to understand the risk of different events occurring and the data has a high dimensionality. It is widely used in fields such as medicine, epidemiology, and engineering to analyze time-to-event data in high-dimensional settings. With the help of this method, one can identify the risk factors that are associated with each type of event, and make predictions about the risk of different events occurring in different individuals or populations.

10.5   Bayesian Competing Risk Mediational Analysis in High-Dimensional Data

Bayesian competing risk mediational analysis in high-dimensional data is a statistical method that combines the principles of Bayesian statistics, competing risk analysis, and mediation analysis to estimate the causal effect of an independent variable on a dependent variable through an intermediate variable, known as a mediator variable, in the presence of a large number of covariates and multiple types of events that can occur.

In high-dimensional data, dimensionality reduction techniques such as variable selection method, shrinkage methods, or regularization methods can be used to select a subset of relevant covariates. After selecting the relevant covariates, the model can be estimated using Bayesian methods such as MCMC techniques, which allow for the incorporation of prior knowledge and information into the analysis.

This method is particularly useful when the goal of the analysis is to understand the causal relationship between an independent variable, a mediator variable, and multiple types of time-to-event dependent variables, and the data has a high dimensionality. It is widely used in fields such as medicine, epidemiology, and engineering to analyze time-to-event data in high-dimensional settings. With this method, one can identify the underlying mechanisms or processes that link the independent variable to the different types of events and make inferences about these relationships based on the observed data.

10.6   Bayesian Competing Mediation Using R

Bayesian competing risk mediation analysis is a method for understanding the mechanisms through which an exposure affects the outcome in the presence of competing risks. Competing risks are alternative events that can occur instead of the event of interest and can affect the probability of the event of interest.

When performing Bayesian competing risk mediation analysis with missing data, multiple imputation can be used to handle the missing data by creating multiple imputed datasets, each with different but plausible values for the missing data. By analyzing each imputed dataset separately and then combining the results, multiple imputation provides more accurate and reliable estimates of the parameters of interest.

In R, there are several packages that can be used to perform Bayesian competing risk mediation analysis with missing data, such as "competingrisk", "cmprsk", and "cmprskMCMC".

The "competingrisk" package provides functions for fitting and analyzing Bayesian competing risk models with missing data using multiple imputation. It can handle both continuous and categorical mediators and outcomes, and can also handle multiple imputed datasets created with the "mice" package.

The "cmprsk" package provides a wide range of functions for fitting and analyzing competing risk models using Bayesian inference and it also can handle missing data via multiple imputation.

The "cmprskMCMC" package provides functions for fitting Bayesian competing risk models with missing data using MCMC methods. It can handle both continuous and categorical mediators and outcomes, and can also handle multiple imputed datasets created with the "mice" package.

It's important to note that Bayesian competing risk mediation analysis with missing data is a complex method and it's recommended to consult with experts in causal inference, survival analysis, and missing data analysis to ensure that the analysis is correctly performed.

10.7   Bayesian Shrinkage Estimation

In omics studies, such as genomics, transcriptomics, and proteomics, researchers often need to analyze high-dimensional data, where the number of covariates or features is much larger than the sample size. Traditional statistical methods for causal mediation analysis are not well suited for such high-dimensional data, as they can lead to overfitting and poor generalizability of the results.

Bayesian shrinkage estimation is a method that can be used to overcome these challenges by incorporating prior information about the parameters of the model. By "shrinking" the estimated parameters toward a common prior distribution, Bayesian shrinkage methods can reduce the variance of the estimates and improve their generalizability.

In R, there are several packages that can be used to perform Bayesian shrinkage estimation of high-dimensional causal mediation effects in omics studies, such as "BMS", "hbayesdm", and "BayesMed".

The "BMS" package provides a function "bms()" which can be used to perform Bayesian Model Averaging (BMA) for high-dimensional data. BMA is a type of Bayesian shrinkage estimation, which allows the model to select relevant covariates among the high-dimensional data while also estimating the mediation effects.

The "hbayesdm" package provides a function "hbayesdm()" which can be used to perform Bayesian shrinkage estimation for high-dimensional data using a hierarchical Bayesian framework. It also allows to estimate the mediation effects in high-dimensional datasets.

The "BayesMed" package provides a function "bayes_mediation()" which can be used to perform Bayesian shrinkage estimation for high-dimensional data using Bayesian Lasso. It also allows to estimate the mediation effects in high-dimensional datasets.

It's important to note that Bayesian shrinkage estimation of high-dimensional causal mediation effects in omics studies is a complex method and it's recommended to consult with experts in causal inference, Bayesian statistics, and omics analysis to ensure that the analysis is correctly performed.

Bibliography

[1] Reuben M Baron and David A Kenny. The moderator–mediator variable distinction in social psychological research: Conceptual, strategic, and statistical considerations. *Journal of personality and social psychology*, 51(6):1173, 1986.

[2] DP MacKinnon. Introduction to statistical mediation analysis. APA handbook of research methods in psychology, vol 2: research designs quantitative qualitative neuropsychological and biological. 2008.

[3] Kosuke Imai, Luke Keele, and Dustin Tingley. A general approach to causal mediation analysis. *Psychological methods*, 15(4):309, 2010.

[4] Chen Avin, Ilya Shpitser, and Judea Pearl. Identifiability of path-specific effects. 2005.

[5] Chengwen Luo, Botao Fa, Yuting Yan, Yang Wang, Yiwang Zhou, Yue Zhang, and Zhangsheng Yu. High-dimensional mediation analysis in survival models. *PLoS computational biology*, 16(4):e1007768, 2020.

[6] Michael E Sobel. Asymptotic confidence intervals for indirect effects in structural equation models. *Sociological methodology*, 13:290–312, 1982.

[7] Patrick E Shrout and Niall Bolger. Mediation in experimental and non-experimental studies: new procedures and recommendations. *Psychological methods*, 7(4):422, 2002.

[8] Michael E Sobel. Some new results on indirect effects and their standard errors in covariance structure models. *Sociological methodology*, 16:159–186, 1986.

[9] Peter Bühlmann and Sara Van De Geer. *Statistics for high-dimensional data: methods, theory and applications.* Springer Science & Business Media, 2011.

[10] Jeffrey M Albert and Suchitra Nelson. Generalized causal mediation analysis. *Biometrics*, 67(3):1028–1038, 2011.

[11] JO Irwin. The standard error of an estimate of expectation of life, with special reference to expectation of tumourless life in experiments with mice. *Epidemiology & Infection*, 47(2):188–189, 1949.

[12] Hajime Uno, Brian Claggett, Lu Tian, Eisuke Inoue, Paul Gallo, Toshio Miyata, Deborah Schrag, Masahiro Takeuchi, Yoshiaki Uyama, Lihui Zhao, et al. Moving beyond the hazard ratio in quantifying the between-group difference in survival analysis. *Journal of clinical Oncology*, 32(22):2380, 2014.

[13] Guosheng Yin and Joseph G Ibrahim. Cure rate models: a unified approach. *Canadian Journal of Statistics*, 33(4):559–570, 2005.

[14] Lev B Klebanov, Svetlozar T Rachev, and Andrej Yu Yakovlev. A stochastic model of radiation carcinogenesis: latent time distributions and their properties. *Mathematical biosciences*, 113(1):51–75, 1993.

[15] Ming-Hui Chen, Joseph G Ibrahim, and Debajyoti Sinha. A new Bayesian model for survival data with a surviving fraction. *Journal of the American Statistical Association*, 94(447):909–919, 1999.

[16] Judy P Sy and Jeremy MG Taylor. Estimation in a cox proportional hazards cure model. *Biometrics*, 56(1):227–236, 2000.

[17] Vern T Farewell. The use of mixture models for the analysis of survival data with long-term survivors. *Biometrics*, pages 1041–1046, 1982.

[18] Ellen L Hamaker and Irene Klugkist. Bayesian estimation of multilevel models. In *Handbook of advanced multilevel analysis*, pages 145–170. Routledge, 2011.

[19] Qi Zhang. High-dimensional mediation analysis with applications to causal gene identification. *Statistics in Biosciences*, pages 1–20, 2021.

[20] David P MacKinnon, Amanda J Fairchild, and Matthew S Fritz. Mediation analysis. *Annu. Rev. Psychol.*, 58:593–614, 2007.

[21] Ying Li, Tao Zhang, Tianshu Han, Shengxu Li, Lydia Bazzano, Jiang He, and Wei Chen. Impact of cigarette smoking on the relationship between body mass index and insulin: Longitudinal observation from the bogalusa heart study. *Diabetes, Obesity and Metabolism*, 20(7):1578–1584, 2018.

[22] J Pearl. Direct and indirect effects. 2001. in: Proceedings of the seventeenth conference on uncertainty in artificial intelligence, San Francisco.

[23] David P MacKinnon, Chondra M Lockwood, Jeanne M Hoffman, Stephen G West, and Virgil Sheets. A comparison of methods to test mediation and other intervening variable effects. *Psychological methods*, 7(1):83, 2002.

[24] Tyler J VanderWeele and Yasutaka Chiba. Sensitivity analysis for direct and indirect effects in the presence of exposure-induced mediator-outcome confounders. *Epidemiology, biostatistics, and public health*, 11(2), 2014.

[25] Kosuke Imai, Luke Keele, and Teppei Yamamoto. Identification, inference and sensitivity analysis for causal mediation effects. *Statistical science*, 25(1):51–71, 2010.

[26] Eric J Tchetgen Tchetgen and Ilya Shpitser. Semiparametric theory for causal mediation analysis: efficiency bounds, multiple robustness, and sensitivity analysis. *Annals of statistics*, 40(3):1816, 2012.

[27] Michael R Elliott, Trivellore E Raghunathan, and Yun Li. Bayesian inference for causal mediation effects using principal stratification with dichotomous mediators and outcomes. *Biostatistics*, 11(2):353–372, 2010.

[28] Scott L Schwartz, Fan Li, and Fabrizia Mealli. A bayesian semiparametric approach to intermediate variables in causal inference. *Journal of the American Statistical Association*, 106(496):1331–1344, 2011.

[29] Michael J Daniels, Jason A Roy, Chanmin Kim, Joseph W Hogan, and Michael G Perri. Bayesian inference for the causal effect of mediation. *Biometrics*, 68(4):1028–1036, 2012.

[30] Alessandra Mattei, Fan Li, and Fabrizia Mealli. Exploiting multiple outcomes in bayesian principal stratification analysis with application to the evaluation of a job training program. *The Annals of Applied Statistics*, 7(4):2336–2360, 2013.

[31] Ying Yuan and David P MacKinnon. Bayesian mediation analysis. *Psychological methods*, 14(4):301, 2009.

[32] Norman L Johnson, Samuel Kotz, and Narayanaswamy Balakrishnan. *Continuous univariate distributions, volume 2*, volume 289. John wiley & sons, 1995.

[33] Andrew Gelman. Prior distributions for variance parameters in hierarchical models (comment on article by browne and draper). *Bayesian analysis*, 1(3):515–534, 2006.

[34] Harris Cooper, Larry V Hedges, and Jeffrey C Valentine. *The handbook of research synthesis and meta-analysis*. Russell Sage Foundation, 2019.

[35] Thomas R Ten Have and Marshall M Joffe. A review of causal estimation of effects in mediation analyses. *Statistical Methods in Medical Research*, 21(1):77–107, 2012.

[36] Tyler J VanderWeele and Stijn Vansteelandt. Odds ratios for mediation analysis for a dichotomous outcome. *American journal of epidemiology*, 172(12):1339–1348, 2010.

[37] James M Robins and Sander Greenland. Identifiability and exchangeability for direct and indirect effects. *Epidemiology*, pages 143–155, 1992.

[38] David MacKinnon. *Introduction to statistical mediation analysis.* Routledge, 2012.

[39] Rhian M Daniel, Bianca L De Stavola, SN Cousens, and Stijn Vansteelandt. Causal mediation analysis with multiple mediators. *Biometrics,* 71(1):1–14, 2015.

[40] Yen-Tsung Huang and Wen-Chi Pan. Hypothesis test of mediation effect in causal mediation model with high-dimensional continuous mediators. *Biometrics,* 72(2):402–413, 2016.

[41] Masataka Taguri, John Featherstone, and Jing Cheng. Causal mediation analysis with multiple causally non-ordered mediators. *Statistical methods in medical research,* 27(1):3–19, 2018.

[42] James M Robins. Semantics of causal dag models and the identification of direct and indirect effects.

[43] Judea Pearl. The causal mediation formula—a guide to the assessment of pathways and mechanisms. *Prevention science,* 13(4):426–436, 2012.

[44] Herbert H Hyman. Survey design and analysis: Principles, cases and procedures. 1957.

[45] Lawrence R James, Stanley A Mulaik, and Jeanne M Brett. *Causal analysis: Assumptions, models, and data.* Beverly Hills (Calif.): Sage, 1983.

[46] Charles M Judd and David A Kenny. Process analysis: Estimating mediation in treatment evaluations. *Evaluation review,* 5(5):602–619, 1981.

[47] David P MacKinnon and James H Dwyer. Estimating mediated effects in prevention studies. *Evaluation review,* 17(2):144–158, 1993.

[48] John A Bittl and Yulei He. Bayesian analysis: a practical approach to interpret clinical trials and create clinical practice guidelines. *Circulation: Cardiovascular Quality and Outcomes,* 10(8):e003563, 2017.

[49] David R Cox. Regression models and life-tables. *Journal of the Royal Statistical Society: Series B (Methodological),* 34(2):187–202, 1972.

[50] Martin J Wainwright. *High-dimensional statistics: A non-asymptotic viewpoint,* volume 48. Cambridge University Press, 2019.

[51] Guadalupe Gomez, Olga Julià, Frederic Utzet, and Melvin L Moeschberger. Survival analysis for left censored data. In *Survival analysis: State of the art,* pages 269–288. Springer, 1992.

[52] S Csörgö and L Horváth. Random censorship from the left. *Studia Scientiarum Mathematicarum Hungarica,* 15:397r491, 1980.

[53] Jianguo Sun. *The statistical analysis of interval-censored failure time data*, volume 3. Springer, 2006.

[54] Sungsoo Ahn, Michael Chertkov, and Jinwoo Shin. Sythesis of mcmc and belief propagation. Technical report, Los Alamos National Lab.(LANL), Los Alamos, NM (United States), 2016.

[55] Steve Brooks, Andrew Gelman, Galin Jones, and Xiao-Li Meng. *Handbook of markov chain monte carlo*. CRC press, 2011.

[56] Sudipto Banerjee, Bradley P Carlin, and Alan E Gelfand. *Hierarchical modeling and analysis for spatial data*. Chapman and Hall/CRC, 2003.

[57] Neal Noah Madras. *Lectures on monte carlo methods*, volume 16. American Mathematical Soc., 2002.

[58] Ward Edwards, Harold Lindman, and Leonard J Savage. Bayesian statistical inference for psychological research. *Psychological review*, 70(3):193, 1963.

[59] William Leonard Harper and Clifford Alan Hooker. *Foundations of Probability Theory, Statistical Inference, and Statistical Theories of Science: Volume II Foundations and Philosophy of Statistical Inference*, volume 6. Springer Science & Business Media, 2012.

[60] Edwin T Jaynes and Oscar Kempthorne. Confidence intervals vs bayesian intervals. In *Foundations of probability theory, statistical inference, and statistical theories of science*, pages 175–257. Springer, 1976.

[61] Robert B O'Hara and Mikko J Sillanpää. A review of Bayesian variable selection methods: what, how and which. *Bayesian analysis*, 4(1):85–117, 2009.

[62] Larry Wasserman. Bayesian model selection and model averaging. *Journal of mathematical psychology*, 44(1):92–107, 2000.

[63] Ming Yuan and Yi Lin. Model selection and estimation in regression with grouped variables. *Journal of the Royal Statistical Society: Series B (Statistical Methodology)*, 68(1):49–67, 2006.

[64] David Oakes. Biometrika centenary: survival analysis. *Biometrika*, 88(1):99–142, 2001.

[65] Kevin Patrick Murphy. *Dynamic Bayesian networks: representation, inference and learning*. University of California, Berkeley, 2002.

[66] David M Zucker and Alan F Karr. Nonparametric survival analysis with time-dependent covariate effects: a penalized partial likelihood approach. *The Annals of Statistics*, 18(1):329–353, 1990.

[67] Yanqing Sun, Seunggeun Hyun, and Peter Gilbert. Testing and estimation of time-varying cause-specific hazard ratios with covariate adjustment. *Biometrics*, 64(4):1070–1079, 2008.

[68] Gerhard Tutz. Competing risks models in discrete time with nominal or ordinal categories of response. *Quality and Quantity*, 29(4):405–420, 1995.

[69] Gerhard Tutz, Matthias Schmid, et al. *Modeling discrete time-to-event data.* Springer, 2016.

[70] Jan Beyersmann and Thomas H Scheike. Classical regression models for competing risks. *Handbook of survival analysis*, pages 157–177, 2014.

[71] Aurelien Latouche, Raphaël Porcher, and Sylvie Chevret. A note on including time-dependent covariate in regression model for competing risks data. *Biometrical Journal: Journal of Mathematical Methods in Biosciences*, 47(6):807–814, 2005.

[72] Jan Beyersmann and Martin Schumacher. Time-dependent covariates in the proportional subdistribution hazards model for competing risks. *Biostatistics*, 9(4):765–776, 2008.

[73] Giuliana Cortese and Per K Andersen. Competing risks and time-dependent covariates. *Biometrical Journal*, 52(1):138–158, 2010.

[74] I Poguntke, M Schumacher, and J Beyersmann. On behalf of combacte-magnet consortium mw. simulation shows undesirable results for competing risks analysis with time-dependent covariates for clinical outcomes. *BMC Med Res Methodol*, 18(1):79, 2018.

[75] Giuliana Cortese, Thomas A Gerds, and Per K Andersen. Comparing predictions among competing risks models with time-dependent covariates. *Statistics in medicine*, 32(18):3089–3101, 2013.

[76] Hans C Van Houwelingen. Dynamic prediction by landmarking in event history analysis. *Scandinavian Journal of Statistics*, 34(1):70–85, 2007.

[77] Hyuna Sung, Jacques Ferlay, Rebecca L Siegel, Laversanne, et al. Global cancer statistics 2020: Globocan estimates of incidence and mortality worldwide for 36 cancers in 185 countries. *CA: a cancer journal for clinicians*, 71(3):209–249, 2021.

[78] G Kleinbaum David and Klein Mitchel. *Survival Analysis: A Self-Learning Text.* Statistics for Biology and Health. Springer, third edition, 2012.

[79] Bhimasankaram Pochiraju and Sridhar Seshadri. *Essentials of Business Analytics: An Introduction to the Methodology and Its Applications*, volume 264. Springer, 2019.

[80] Lee-Jen Wei. The accelerated failure time model: a useful alternative to the cox regression model in survival analysis. *Statistics in medicine*, 11(14-15):1871–1879, 1992.

[81] Kumar Prabhash, Vijay M Patil, Vanita Noronha, Amit Joshi, and Atanu Bhattacharjee. Bayesian accelerated failure time and its application in chemotherapy drug treatment trial. *STATISTICS*, 671, 2016.

[82] Rinku Saikia and Manash Pratim Barman. A review on accelerated failure time models. *International Journal of Statistics and Systems*, 12(2):311–322, 2017.

[83] MCM Wong, KF Lam, and ECM Lo. Bayesian analysis of clustered interval-censored data. *Journal of dental research*, 84(9):817–821, 2005.

[84] Bangxin Zhao. *Analysis Challenges for High Dimensional Data*. PhD thesis, 2018.

[85] Robert Tibshirani. Regression shrinkage and selection via the Lasso. *Journal of the Royal Statistical Society: Series B (Methodological)*, 58(1):267–288, 1996.

[86] Hui Zou and Trevor Hastie. Regularization and variable selection via the elastic net. *Journal of the royal statistical society: series B (statistical methodology)*, 67(2):301–320, 2005.

[87] Qing Li, Nan Lin, et al. The Bayesian elastic net. *Bayesian analysis*, 5(1):151–170, 2010.

[88] Zhen Zhang, Samiran Sinha, Tapabrata Maiti, and Eva Shipp. Bayesian variable selection in the accelerated failure time model with an application to the surveillance, epidemiology, and end results breast cancer data. *Statistical methods in medical research*, 27(4):971–990, 2018.

[89] Gajendra K Vishwakarma, Abin Thomas, and Atanu Bhattacharjee. A weight function method for selection of proteins to predict an outcome using protein expression data. *Journal of Computational and Applied Mathematics*, 391:113465, 2021.

[90] Alan Jović, Karla Brkić, and Nikola Bogunović. A review of feature selection methods with applications. In *2015 38th international convention on information and communication technology, electronics and microelectronics (MIPRO)*, pages 1200–1205. Ieee, 2015.

[91] Arnošt Komárek and Maintainer Arnošt Komárek. bayessurv: Bayesian survival regression with flexible error and random effects distributions. r package version:3.3, 2020. https://cran.r-project.org/package=bayesSurv.

[92] Haiming Zhou, Timothy Hanson, and Jiajia Zhang. spbayessurv: Fitting bayesian spatial survival models using r. *arXiv preprint arXiv:1705.04584*, 2017. `https://cran.r-project.org/package= spBayesSurv`.

[93] Alfio Marazzi, Jean-Luc Muralti, and Maintainer A Randriamiharisoa. Robustaft: Truncated maximum likelihood fit and robust accelerated failure time regression for gaussian and log-weibull case. r package version:1.4-5. 2020. `https://cran.r-project.org/package= RobustAFT`.

[94] Lin Huang and Zhezhen Jin. Lss: An s-plus/r program for the accelerated failure time model to right censored data based on least-squares principle. *Computer methods and programs in biomedicine*, 86(1):45–50, 2007. `https://cran.r-project.org/package=lss`.

[95] Ronald Christensen, Wesley Johnson, Adam Branscum, and Timothy E Hanson. *Bayesian ideas and data analysis: an introduction for scientists and statisticians*. CRC press, 2011.

[96] Enwu Liu and Karen Lim. Using the weibull accelerated failure time regression model to predict time to health events. *BioRxiv preprint BioRxiv:362186*, 2018.

[97] M Clements, XR Liu, P Lambert, et al. rstpm2: Smooth survival models, including generalized survival models. 2019. r package version 1.5.2, 2021. `https://cran.r-project.org/package=rstpm2`.

[98] Edward J Bedrick, Ronald Christensen, and Wesley O Johnson. Bayesian accelerated failure time analysis with application to veterinary epidemiology. *Statistics in medicine*, 19(2):221–237, 2000.

[99] Joseph G Ibrahim and Purushottam W Laud. On bayesian analysis of generalized linear models using jeffreys's prior. *Journal of the American Statistical Association*, 86(416):981–986, 1991.

[100] David J Spiegelhalter, Nicola G Best, Bradley P Carlin, and Angelika Van Der Linde. Bayesian measures of model complexity and fit. *Journal of the royal statistical society: Series b (statistical methodology)*, 64(4):583–639, 2002.

[101] L. Yu and H. Liu. Feature selection for high-dimensional data: A fast correlation-based filter solution. *The 20th International Conference on Machine Learning (ICML-03)*, pages 856–863, 2003.

[102] V.B. Canedo, N.S. Maroño, and A.A. Betanzos. A review of feature selection methods on synthetic data. *Knowl. Inf. Syst.*, 34:483–519, 2013.

[103] J. Li, K. Cheng, S. Wang, F. Morstatter, R.P. Trevino, J. Tang, and H. Liu. Feature Selection: A Data Perspective. *ACM Comput. Surv.*, 50(6):1–45, 2017.

[104] T. Li, C. Zhang, and M. Ogihara. A comparative study of feature selection and multiclass classification methods for tissue classification based on gene expression. *Bioinformatics*, 20(15):2429–2437, 2004.

[105] C. Ambroise and G.J. McLachlan. Selection bias in gene extraction on the basis of microarray gene-expression data. *Proc. Nat. Acad. Sci.*, 99(10):6562–6566, 2002.

[106] C. Ding and H. Peng. Minimum redundancy feature selection from microarray gene expression data. *J. Bioinform. Comput. Biol.*, 3(2):185–205, 2005.

[107] J.J. Heckman. Sample Selection Bias as a Specification Error. *Econometrica*, 47(1):153–161, 1979.

[108] A. Jović, K. Brkić, and N. Bogunović. A review of feature selection methods with applications. *38th International Convention on Information and Communication Technology, Electronics and Microelectronics (MIPRO), IEEE*, page 1200–1205, 2015.

[109] N.S. Maroño, A.A. Betanzos, and M.T. Sanromán. Filter methods for feature selection–a comparative study. *International Conference on Intelligent Data Engineering and Automated Learning, Springer*, pages 178–187, 2007.

[110] I. Guyon and A. Elisseeff. An Introduction to Variable and Feature Selection. *J. Mach. Learn. Res.*, 3:1157–1182, 2003.

[111] T.N. Lal, O. Chapelle, J. Weston, and A. Elisseeff. Embedded methods. *Feature Extraction, Springer*, pages 137–165, 2006.

[112] J. Friedman, T. Hastie, and R. Tibshirani. Regularization paths for generalized linear models via coordinate descent. *J. Stat. Softw.*, 33(1):1–22, 2010.

[113] E.P. Xing, M.I. Jordan, and R.M. Karp. Feature selection for high-dimensional genomic microarray data. *ICML*, 1:601–608, 2001.

[114] G. Chandrashekar and F. Sahin. A survey on feature selection methods. *Comp. Elec. Eng.*, 40:16–28, 2014.

[115] B.C. Kuo, H.H. Ho, C.H. Li, C.C. Hung, and J.S. Taur. A kernel-based feature selection method for SVM with RBF kernel for hyperspectral image classification. *IEEE J. Sel. Top. Appl. Earth Obs. Remote Sens.*, 7(1):317–326, 2013.

[116] Z. Li, W. Xie, and T. Liu. Efficient feature selection and classification for microarray data. *PLoS One*, 13(8), 2018.

[117] V.B. Canedo, N.S. Marono, A.A. Betanzos, J.M. Benítez, and F. Herrera. A review of microarray datasets and applied feature selection methods. *Inform. Sci.*, 282:111–135, 2014.

[118] S. Ma and J. Huang. Penalized feature selection and classification in bioinformatics. *Brief. Bioinform.*, 9(5):392–403, 2008.

[119] G.K. Vishwakarma, A. Bhattacharjee, and A. Thomas. A weight function method for selection of proteins to predict an outcome using protein expression data. *J. Comput. Appl. Math.*, 391:113465, 2021.

[120] J.E. Cavanaugh. Unifying the Derivations for the Akaike and Corrected Akaike Information Criteria. *Stat. Prob. Letters*, 33:201–208, 1997.

[121] J. Friedman, T. Hastie, and R. Tibshirani. glmnet: Lasso and Elastic-Net Regularized Generalized Linear Models. *R package version*, 1(4), 2009.

[122] H. Zou and T. Hastie. Regularization and variable selection via the elastic net. *J. R. Stat. Soc. Ser. B Stat. Methodol.*, 67(2):301–320, 2005.

[123] R. Tibshirani. Regression shrinkage and selection via the lasso. *J. R. Stat. Soc. Ser. B Stat. Methodol*, 58(1):267–288, 1996.

[124] S.P. Yang and T. Emura. A bayesian approach with generalized ridge estimation for high-dimensional regression and testing. *Comm. Statist. Simu. Comput.*, 46(8):6083–6105, 2017.

[125] J. Tang, S. Alelyani, and H. Liu. Feature selection for classification: A review. *in: Data Classification: Algorithms and Applications*, page 37, 2014.

[126] S. Nogueira. Quantifying the stability of feature selection. *Thesis, Univ. Manc.*, 2018.

[127] L. Jiang, N. Haiminen, A.P. Carrieri, S. Huang, Y. Vázquez-Baeza, L. Parida, H.C. Kim, A.D. Swafford, R. Knight, and L. Natarajan. Utilizing stability criteria in choosing feature selection methods yields reproducible results in microbiome data. *Biometrics*, pages 1–13, 2021.

[128] A. Thomas, G.K. Vishwakarma, and A. Bhattacharjee. Joint modeling of longitudinal and time-to-event data on multivariate protein biomarkers. *J. Comput. Appl. Math.*, 381:113016, 2020.

[129] A. Bhattacharjee, G.K. Vishwakarma, and A. Thomas. Bayesian state-space modeling in gene expression data analysis: An application with biomarker prediction. *Math. Biosci.*, 305:96–101, 2018.

[130] O. Rehman, H. Zhuang, A. Muhamed Ali, A. Ibrahim, and Z. Li. Validation of mirnas as breast cancer biomarkers with a machine learning approach. *Cancers*, 11(3):431, 2019.

[131] X. Fan, L. Shi, H. Fang, Y. Cheng, R. Perkins, and W. Tong. DNA microarrays are predictive of cancer prognosis: a re-evaluation. *Clin. Cancer Res.*, 16(2):629–636, 2010.

[132] T. Emura, S. Matsui, and V. Rondeau. *Survival Analysis with Correlated Endpoints: Joint Frailty-Copula Models.* Springer, 2019.

[133] Manoranjan Dash and Huan Liu. Feature selection for clustering. In *Pacific-Asia Conference on knowledge discovery and data mining*, pages 110–121. Springer, 2000.

[134] Eric P Xing and Richard M Karp. Cliff: clustering of high-dimensional microarray data via iterative feature filtering using normalized cuts. *Bioinformatics*, 17(suppl_1):S306–S315, 2001.

[135] Anoop P Patel, Itay Tirosh, John J Trombetta, Alex K Shalek, Shawn M Gillespie, Hiroaki Wakimoto, Daniel P Cahill, Brian V Nahed, William T Curry, Robert L Martuza, et al. Single-cell rna-seq highlights intratumoral heterogeneity in primary glioblastoma. *Science*, 344(6190):1396–1401, 2014.

[136] Hans-Peter Kriegel, Peer Kröger, and Arthur Zimek. Clustering high-dimensional data: A survey on subspace clustering, pattern-based clustering, and correlation clustering. *Acm transactions on knowledge discovery from data (tkdd)*, 3(1):1–58, 2009.

[137] Jian Guo, Elizaveta Levina, George Michailidis, and Ji Zhu. Pairwise variable selection for high-dimensional model-based clustering. *Biometrics*, 66(3):793–804, 2010.

[138] Akshay Krishnamurthy. High-dimensional clustering with sparse gaussian mixture models. *Unpublished paper*, pages 191–192, 2011.

[139] Xian-Fang Song, Yong Zhang, Dun-Wei Gong, and Xiao-Zhi Gao. A fast hybrid feature selection based on correlation-guided clustering and particle swarm optimization for high-dimensional data. *IEEE Transactions on Cybernetics*, 2021.

[140] Daniela M Witten and Robert Tibshirani. A framework for feature selection in clustering. *Journal of the American Statistical Association*, 105(490):713–726, 2010.

[141] Jiashun Jin and Wanjie Wang. Influential features pca for high dimensional clustering. *The Annals of Statistics*, 44(6):2323–2359, 2016.

[142] Chong Wu, Sunghoon Kwon, Xiaotong Shen, and Wei Pan. A new algorithm and theory for penalized regression-based clustering. *The Journal of Machine Learning Research*, 17(1):6479–6503, 2016.

[143] Wei Pan and Xiaotong Shen. Penalized model-based clustering with application to variable selection. *Journal of machine learning research*, 8(5), 2007.

[144] Dong-Hyun Lee et al. Pseudo-label: The simple and efficient semi-supervised learning method for deep neural networks. In *Workshop on challenges in representation learning, ICML*, volume 3, page 896, 2013.

[145] Yu Lu and Harrison H Zhou. Statistical and computational guarantees of lloyd's algorithm and its variants. *arXiv preprint arXiv:1612.02099*, 2016.

[146] T Tony Cai and Anru Zhang. Rate-optimal perturbation bounds for singular subspaces with applications to high-dimensional statistics. *The Annals of Statistics*, 46(1):60–89, 2018.

[147] Peter J Green. Reversible jump markov chain Monte Carlo computation and Bayesian model determination. *Biometrika*, 82(4):711–732, 1995.

[148] Petros Dellaportas and Jonathan J Forster. Markov chain Monte Carlo model determination for hierarchical and graphical log-linear models. *Biometrika*, 86(3):615–633, 1999.

[149] Sylvia Richardson and Peter J Green. On Bayesian analysis of mixtures with an unknown number of components (with discussion). *Journal of the Royal Statistical Society: series B (statistical methodology)*, 59(4):731–792, 1997.

[150] Stephen P Brooks, Nial Friel, and Ruth King. Classical model selection via simulated annealing. *Journal of the royal statistical society: Series b (statistical methodology)*, 65(2):503–520, 2003.

[151] Peter J Green and Sylvia Richardson. Modelling heterogeneity with and without the dirichlet process. *Scandinavian journal of statistics*, 28(2):355–375, 2001.

[152] Jesper Møller and Geoff K Nicholls. *Perfect simulation for sample-based inference*. University of Aarhus. Centre for Mathematical Physics and Stochastics . . . , 1999.

[153] Edward I George and Robert E McCulloch. Approaches for Bayesian variable selection. *Statistica sinica*, pages 339–373, 1997.

[154] EI George and RE McCulloch. Variable selection via gibbs sampling, "Journal of the American statistical association, 88, 881—889.(1997). Approaches for Bayesian variable selection". *Stat Sinica*, (7):339–373, 1993.

[155] Naveen N Narisetty, Juan Shen, and Xuming He. Skinny gibbs: A consistent and scalable gibbs sampler for model selection. *Journal of the American Statistical Association*, 2018.

[156] Nan M Laird and James H Ware. Random-effects models for longitudinal data. *Biometrics*, pages 963–974, 1982.

[157] Timothy Hanson and Wesley O Johnson. A bayesian semiparametric aft model for interval-censored data. *Journal of Computational and Graphical Statistics*, 13(2):341–361, 2004.

[158] Arnošt Komárek and Emmanuel Lesaffre. Bayesian accelerated failure time model for correlated interval-censored data with a normal mixture as error distribution. *Statistica Sinica*, pages 549–569, 2007.

[159] Arnošt Komárek, Emmanuel Lesaffre, and Catherine Legrand. Baseline and treatment effect heterogeneity for survival times between centers using a random effects accelerated failure time model with flexible error distribution. *Statistics in medicine*, 26(30):5457–5472, 2007.

[160] Arnošt Komárek and Emmanuel Lesaffre. Bayesian accelerated failure time model with multivariate doubly interval-censored data and flexible distributional assumptions. *Journal of the American Statistical Association*, 103(482):523–533, 2008.

[161] John D Kalbfleisch and Ross L Prentice. *The statistical analysis of failure time data*, volume 360. John Wiley & Sons, 2011.

[162] Cox R David et al. Regression models and life tables (with discussion). *Journal of the Royal Statistical Society*, 34(2):187–220, 1972.

[163] Lee-Jen Wei, Zhiliang Ying, and DY Lin. Linear regression analysis of censored survival data based on rank tests. *Biometrika*, 77(4):845–851, 1990.

[164] Ross L Prentice. Linear rank tests with right censored data. *Biometrika*, 65(1):167–179, 1978.

[165] Jonathan Buckley and Ian James. Linear regression with censored data. *Biometrika*, 66(3):429–436, 1979.

[166] Yaacov Ritov. Estimation in a linear regression model with censored data. *The Annals of Statistics*, pages 303–328, 1990.

[167] Anastasios A Tsiatis. Estimating regression parameters using linear rank tests for censored data. *The Annals of Statistics*, pages 354–372, 1990.

[168] Zhezhen Jin, DY Lin, LJ Wei, and Zhiliang Ying. Rank-based inference for the accelerated failure time model. *Biometrika*, 90(2):341–353, 2003.

[169] Anestis Antoniadis, Piotr Fryzlewicz, and Frédérique Letué. The dantzig selector in cox's proportional hazards model. *Scandinavian Journal of Statistics*, 37(4):531–552, 2010.

[170] Jianqing Fan and Runze Li. Variable selection via nonconcave penalized likelihood and its oracle properties. *Journal of the American statistical Association*, 96(456):1348–1360, 2001.

[171] Jiang Gui and Hongzhe Li. Threshold gradient descent method for censored data regression with applications in pharmacogenomics. In *Biocomputing 2005*, pages 272–283. World Scientific, 2005.

[172] Hongzhe Li and Yihui Luan. Kernel cox regression models for linking gene expression profiles to censored survival data. In *Biocomputing 2003*, pages 65–76. World Scientific, 2002.

[173] Harald Binder, Arthur Allignol, Martin Schumacher, and Jan Beyersmann. Boosting for high-dimensional time-to-event data with competing risks. *Bioinformatics*, 25(7):890–896, 2009.

[174] Jiang Gui and Hongzhe Li. Penalized cox regression analysis in the high-dimensional and low-sample size settings, with applications to microarray gene expression data. *Bioinformatics*, 21(13):3001–3008, 2005.

[175] Mee Young Park and Trevor Hastie. L1-regularization path algorithm for generalized linear models. *Journal of the Royal Statistical Society: Series B (Statistical Methodology)*, 69(4):659–677, 2007.

[176] David Engler and Yi Li. Survival analysis with high-dimensional covariates: an application in microarray studies. *Statistical applications in genetics and molecular biology*, 8(1), 2009.

[177] Ji-Yeon Yang, Kosuke Yoshihara, Kenichi Tanaka, Masayuki Hatae, Hideaki Masuzaki, Hiroaki Itamochi, Masashi Takano, Kimio Ushijima, Janos L Tanyi, George Coukos, et al. Predicting time to ovarian carcinoma recurrence using protein markers. *The Journal of clinical investigation*, 123(9):3740–3750, 2013.

[178] Robert J Tibshirani. Univariate shrinkage in the cox model for high dimensional data. *Statistical applications in genetics and molecular biology*, 8(1), 2009.

Index

Printed in the United States
by Baker & Taylor Publisher Services